Geology and hydrogeology of the Teays-Mahomet Bedrock Valley System

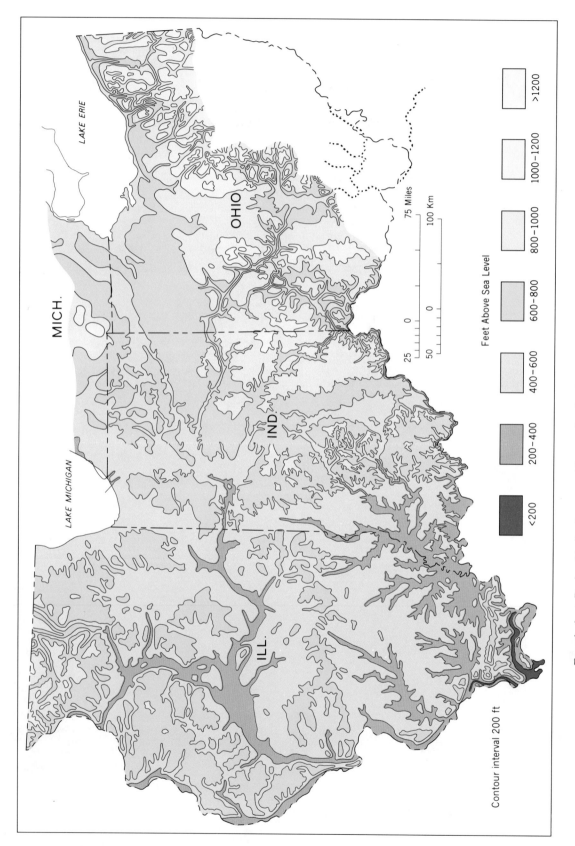

Frontispiece. Regional bedrock topographic map showing Teays Valley system from West Virginia to Illinois. Compiled by Henry H. Gray (this volume) from the following sources: Ohio, Cummins (1959); Indiana, Gray (1982); Michigan, Grant and Pringle (1943); Illinois, Horberg (1950). Outside contoured area, dotted lines show surface expression of Teays Valley system, from Tight (1903). See Gray (this volume) for references.

Geology and hydrogeology of the Teays-Mahomet Bedrock Valley Systems

Edited by

Wilton N. Melhorn
Department of Earth and Atmospheric Sciences
Purdue University
West Lafayette, Indiana 47907

John P. Kempton
Illinois State Geological Survey
Natural Resources Building
615 East Peabody Drive
Champaign, Illinois 61820

SPECIAL PAPER
258
1991

© 1991 The Geological Society of America, Inc.
All rights reserved.

All materials subject to this copyright and included
in this volume may be photocopied for the noncommercial
purpose of scientific or educational advancement.

Published by The Geological Society of America, Inc.
3300 Penrose Place, P.O. Box 9140, Boulder, Colorado 80301

Printed in U.S.A.

GSA Books Science Editor Richard A. Hoppin

Library of Congress Cataloging-in-Publication Data

Geology and hydrogeology of the Teays-Mahomet bedrock valley system /
  edited by Wilton N. Melhorn. John P. Kempton.
      p.   cm.—(Special paper ; 258)
  Includes bibliographical references.
  ISBN 0-8137-2258-6
  1. Geology—Middle West. 2. Water, Underground—Middle West
  3. Glacial epoch—Middle West. I. Melhorn, Wilton Newton, 1920-
. II. Kempton, John P. (John Paul) III. Title: Teays-Mahomet
bedrock valley system. IV. Series: Special papers (Geological
Society of America) ; 258.
QE78.7.G46   1991
557.7—dc20   91-11131
                                                         CIP

Cover photo: Enlargement of a photo of an ERTS image shows the Ohio segment of the classic Teays Valley of Tight. Abandoned Teays Valley, generally paralleling the Scioto River Valley to the east, is partially coincident with the Scioto to the north and generally coincident with the Ohio River to the southeast. Chillicothe, Ohio, on the Scioto River, is at the northern edge of the photo area; Portsmouth, Ohio, is on the Ohio River between the Scioto and the Teays Valley and about 45 miles to the south of Chillicothe. (For further orientation, see frontispiece, Figs. 1 and 2 in Goldthwait, Fig. 1 in Bonnett and others, and Fig. 1 in Bigham and others, this volume.) Color slide courtesy of R. B. Bonnett, Marshall University.

10 9 8 7 6 5 4 3 2 1

Contents

Foreword vii

Introduction 1
    Wilton N. Melhorn and John P. Kempton

The Teays Valley problem; A historical perspective 3
    Richard P. Goldthwait

*A paleomagnetic study of the early Pleistocene Minford
    Silt member, Teays Formation, West Virginia* 9
    Richard B. Bonnett, Hallan C. Noltimier, and
    Dewey D. Sanderson

*Lithology and general stratigraphy of Quaternary sediments
    in a section of the Teays River Valley of southern Ohio* 19
    J. M. Bigham, N. E. Smeck, L. D. Norton, G. F. Hall,
    and M. L. Thompson

The Old Kentucky River; A major tributary to the Teays River 29
    James T. Teller and Richard P. Goldthwait

*Origin and history of the Teays drainage system: The view
    from midstream* 43
    Henry H. Gray

*The Lafayette Bedrock Valley System of Indiana;
    Concept, form, and fill stratigraphy* 51
    N. K. Bleuer

*Aquifer systems of the buried Marion-Mahomet trunk valley
    (Lafayette Bedrock Valley System) of Indiana* 79
    N. K. Bleuer, W. N. Melhorn, W. J. Steen, and T. M. Bruns

Mahomet Bedrock Valley in east-central Illinois; Topography, glacial drift stratigraphy, and hydrogeology 91
   John P. Kempton, W. Hilton Johnson, Paul C. Heigold, and
   Keros Cartwright

The Teays System; A summary 125
   Wilton N. Melhorn and John P. Kempton

Foreword

Relatively few published works in the geosciences have appeal for workers in both basic and applied fields. This compendium of papers is one such effort. It is an outgrowth of a symposium held at Indianapolis during the Annual Meeting of the Geological Society of America in 1983, and was sponsored by both the Society's Division of Hydrogeology and the Division of Quaternary Geology and Geomorphology. It contains a fascinating detective story about a geologic entity that has been simultaneously overstated and understudied. For the most part, the authors of this volume challenge old adages. Yet throughout, even though previous works are challenged, previous workers are not impugned.

The Teays Bedrock Valley System is the epitome of a drainage system affected by continental glaciation. It is a patchwork of stream segments, extending from the headwaters of the New River in western North Carolina to the point in central Illinois where it merges with the Illinois River System. Likewise, the study of the Teays System is also a patchwork, reminiscent of the blind man's description of the elephant. Indeed, the size, diversity, and in places, the obscurity of the system dictate that initial study, at least, must be confined to relatively small parts of the whole. Even the comprehensive early study by W. G. Tight failed to recognize the full extent of the system. In fact, not until the 1940s was the continuity of the system across the glaciated portions of northwestern Ohio, central Indiana, and east-central Illinois firmly established. This was the era of the Second World War, when water-supply needs for industry were paramount. Thus, the Teays-Mahomet System became an issue with considerable resource impact, not just an abstract scientific debate. The system was investigated as a resource, although scientific curiosity as to the why and where of the system still was predominant.

At this time, it was tacitly assumed that the surface of the bedrock—buried by glacial drift throughout the north-central United States, and the landscape presently visible in areas marginal to the glaciated terrain, including the Driftless Area of southwestern Wisconsin—was preglacial in age. Hence, the valleys on this bedrock surface, such as the Teays, must also be considered preglacial, and as such, must represent an integrated drainage network. The concept of this Teays-Mahomet System was almost romantic—it appealed to the keep-it-simple, big-picture primordial urge in the minds of classical geologists. As is so often the case in scientific matters, once in print, always the norm.

This dogma was challenged by John Frye, who pointed out that much erosion of the bedrock occurred during and subsequent to the numerous episodes of ice advance, and that glacio-isostatic effects on stream behavior, forebulge as well as depression, though difficult to recognize, cannot be ignored. In short, the history of the bedrock surface is complex, and the evidence upon which it rests is often obscure and difficult to decipher.

The present volume offers abundant confirmation of these insights. Most of the authors of this volume challenge the concept of a single, preglacial system. The evidence is overwhelming: (1) the Teays System is indeed a patchwork of individual segments; (2) it may never have functioned as an integrated drainage network; and (3) its history is complex, obscure, and difficult to decipher. In a sense, this volume brings the study of the Teays System full circle. Early studies focused on the parts and did not see the whole. By 1950, the focus was generally on the whole, and the system was considered to be an integrated, preglacial drainage network. This volume, though it shows the Teays to be a "system," a collection of interconnected and interdependent parts, clearly demonstrates that the whole is actually the sum of many quite dissimilar parts. Furthermore, the studies reported in this volume confirm that the Teays Bedrock Valley System is the most intensively investigated buried valley in the glaciated north-central United States, and assure that it will remain the point of departure for investigation of similar valleys within the region.

Current studies, represented by this volume, are again characterized by the dual activities of basic and applied geoscientists. As a result of efforts to develop a better understanding of ground-water availability and flow patterns, and thus to help ensure protection of ground-water quality, a more realistic picture of the system is emerging. New tools are being introduced, including geophysics and paleomagnetics. New concepts of glacial stratigraphy and cause-and-effect relations are being utilized. As pointed out by Melhorn and Kempton, ultimate answers are likely to be derived only when the complications of such probable events as bedrock forebulges associated with continental glaciation can be recognized and their impact assessed. In the meantime, as demonstrated by this volume, capable researchers continue to do the necessary detective work. Perhaps the most lasting value of this volume may lie in the stimulus that it will provide for the study of other, equally beguiling, buried valleys.

Both of us were weaned into the era when the system was intact. To recognize now that the system equals "Systems" is both educational and fascinating. Whereas this is not now an issue of widespread public concern, it could be in the future, given the nature of the story and the resource base located within the systems. Laypersons increasingly depend on geoscientists to help them understand and utilize the physical environment and resource base. Thus, it is important for geoscientists to understand and communicate. The authors of this volume do both.

Dr. Richard C. Anderson
Professor of Geology
Augustana College
Rock Island, Illinois 61201

Dr. David A. Stephenson
President
Geoscience Reviews, Inc.
8669 East San Alberto Drive
Scottsdale, Arizona 85258

July 28, 1989

Introduction

Wilton N. Melhorn
Department of Earth and Atmospheric Sciences, Purdue University, West Lafayette, Indiana 47907

John P. Kempton
Illinois State Geological Survey, Natural Resources Building, 615 East Peabody Drive, Champaign, Illinois 61820

Most Midwest Friends of the Pleistocene (FOP) field conferences provide an opportunity for "friends" to discuss a variety of issues. From these discussions may occasionally emerge "great" ideas. Such was the case in the spring of 1981 during the 29th Annual Midwest FOP field conference held in the northeastern lower peninsula of Michigan. Seated for lunch near a bedrock waterfall, the co-editors of this volume were exchanging pleasantries when, in a moment of enlightenment, Melhorn burst forth with, "we ought to have a symposium on the Teays in 1983 at Indy!" In a moment of weakness, Kempton agreed, and a symposium was organized for the Annual Meeting of the Geological Society of America at Indianapolis in 1983. Since that time, the successful, well-attended symposium is just a memory, and the symposium volume, supported by manuscripts supplied by most of the speakers, is at last seeing the light of day. To the authors who contributed timely manuscripts we offer a big *thanks* for your patience; to the readers, we believe that, despite the long lapse from meeting to publication, the material is worth the wait. In fact, some updating was done where appropriate during the review process.

The Teays Valley has been a part of the editors' professional heritage for many years. Kempton's interest began while filing topographic maps as a geology department assistant at Denison University, when he noted on one of the new 7½-min topographic quadrangle maps a large, strange valley containing no river of any significance. His first "independent" study determined this valley to be a segment of the Teays Valley. A few years later, as a graduate student under Goldthwait at The Ohio State University, he encountered abandoned segments of Teays-age tributaries during master's thesis work on outwash terraces in the Hocking River Valley, and later during terrace work in the Scioto Valley, he met head-on with the main Teays Valley. Subsequent encounters with the generally accepted lower segment of the Teays, the Mahomet Valley in Illinois, during the last 30 years while with the Illinois State Geological Survey have been frequent in helping answer questions about groundwater availability, and have continued to fan his research interest.

Melhorn, during more than three decades at Purdue University, constantly has engaged in disabusing the local folklore that a "buried river full of water" lies directly beneath the Lafayette community. The symposium idea at the waterfall was the result of a preoccupation with ongoing studies of examining the late glacial history of the middle Wabash Valley that culminated in the organization of the 30th Annual Midwest Friends of the Pleistocene field conference in that region in 1983.

More than 40 years have elapsed since Leland Horberg summarized available data to show possible preglacial drainage systems in the central United States. Among these systems, the Teays-Mahomet is the classic prototype of a large "preglacial valley" or "buried valley." It receives mention in some introductory physical-historical geology texts, and commonly is discussed in general glacial geology, geomorphology, and hydrogeology texts. Indeed, it has received international recognition, and has also been the subject of numerous stories in the media.

As with many subjects of scientific interest, study of the Teays-Mahomet System has proceeded episodically for almost 20 years. Beginning in the late 1950s, after Horberg's death and publication of the Ohio studies of Norris, Goldthwait, and others, the Teays-Mahomet System received little attention. In recent years, however, new geophysical programs, improvements in sedimentological techniques and analysis, renewed stratigraphic studies, and reexamination of the drainage history in southern and southwestern Ohio, West Virginia, Kentucky, and southeastern Indiana have been coupled with rather extensive drilling programs associated with hydrogeological investigations in central and east-central Illinois and across central Indiana. These have provided a wealth of new data, which require reinterpretation of the late Tertiary and Pleistocene history of this valley system and reassessment of the local regional potential of the valley-fill sediments as important aquifers. Some of this information has been disseminated as oral presentations and theses, as open-file reports, or in publications of the U.S. Geological Survey, state geological surveys, and Geological Society of America publications. Yet, assuredly, there is no up-to-date broad regional synthesis volume

Melhorn, W. N., and Kempton, J. P., 1991, Introduction, *in* Melhorn, W. N., and Kempton, J. P., eds., Geology and hydrogeology of the Teays-Mahomet Bedrock Valley System: Boulder, Colorado, Geological Society of America Special Paper 258.

wherein recent findings are combined with earlier work to create a modern analysis and reinterpretation of the bedrock topography, alluvial fill history, stratigraphy, and hydrology of the Teays-Mahomet System.

The heart of this buried valley system is in Indiana; the presumed master channel transects the state, and perhaps the "final solution" to many key questions concerning the history of the valley and regional groundwater resources may be found there. Therefore, the theme was appropriate and timely for presentation at Indianapolis, and the results are incorporated into this symposium volume. We believe this volume is a significant contribution to the state of knowledge about an important geologic feature.

The makeup of this volume differs somewhat from the original schedule of oral presentation. For example, three oral papers on the Mahomet System in Illinois are combined into a single paper in order to eliminate redundancy of illustration and flow within the volume, and an additional paper on aquifer systems of the Marion-Mahomet valley in Indiana has been added to strengthen the hydrogeological portion of the work. Furthermore, as the available older and classic data, published and unpublished, were not recorded in the metric system, and because many of the maps prepared for this project will be utilized for public information, metric conversions may appear only by scale in the figures and in a few places within the text of some papers. This procedure, where used, avoids the distracting repetition of providing both metric and English standard values, retains the purity of classic measurements and descriptions, and obviates the chance of conversion error. In general, we have allowed the author free choice as long as each chapter is internally consistent.

We are grateful for the joint sponsorship of the symposium by the Quaternary Geology and Geomorphology and the Hydrogeology Divisions of the Geological Society of America.

The Teays Valley problem; A historical perspective

Richard P. Goldthwait
P.O. Box 656, Anna Maria, Florida 34216

ABSTRACT

The exposed and abandoned Teays Valley has been recognized and studied in south-central Ohio for a century and a half. By 1900, the upper reaches had been traced by bedrock strath up present deeper drainages through West Virginia into Virginia. By 1945, the 1.5-mi-wide main valley had been traced downstream by water/oil wells, under several glacial drifts, northward and westward from Chillicothe, Ohio, to Indiana. The average gradient in this reach is northwest 0.9 ft/mi. Horberg and others carried this valley westward to the Mahomet buried drainage of Illinois.

Discontinuities such as in Madison County, Ohio, have been explored by geophysical profiles and test wells and show a marked narrowing into a canyon cut through resistant dolomites. As late as the 1970s, the largest tributary, which drains most of eastern Kentucky northward and on either side of Cincinnati, has been added.

More and more basal floodplain sands/gravels and northward-inclined crossbeds are found in all these meandering valleys beyond the glacial limit. Near the limit these deposits are covered with lacustrine clay rhythmites, which grade to silty and then into sandy outwash containing glacially derived northern clasts. Clearly, glaciation dammed the valley system westward out of Ohio. This blockage was early Pleistocene, certainly pre-Illinoian, because clays are magnetically reversed. Other blockages took place southwestward from Cincinnati at a later date. The exposed Minford Clays in south-central Ohio and western West Virginia must have filled to near the present 900-ft contour, because the former, dendritic Teays drainage is criss-crossed by an aimless, superimposed drainage that postdates the Deep Stage.

DISCOVERY

A century and a half ago, when the theory of continental glaciation had just been conceived, one astute geologist, J. Hildreth, recognized that the complex criss-crossing valleys in southern Ohio and northern Kentucky meant that "Great changes have evidently been made in the direction of all our water courses" . . . (Hildreth, 1838). Others, like Drake of Cincinnati in the early 19th century, observed intersecting valleys with fertile farms high (and hanging) above present rivers (Drake, 1825). Throughout the rest of the 19th century, geologists puzzled over the complications of the Ohio River Valley, especially upstream from Cincinnati: why did the bedrock valley widen in some places and narrow elsewhere? These observations led to hypotheses that the land had flexed upward, first in one place, then in another, to capture crossing drainage systems. Thus a formerly northwestward-flowing drainage took a more direct, southwestward shortcut to the Tertiary sea in the upper Mississippi Embayment. By the end of the 19th century the reality of continental glaciation was fully demonstrated and generally accepted; of the two major glaciations then recognized, the earlier (Illinoian), had crossed the Ohio River (from Ripley, Ohio, to Carrollton, Kentucky), dammed the waters temporarily, and caused deposition of some lacustrine clays near Cincinnati. Both glaciations locally caused many tributary extensions (captures), diversions, and temporary pondings along the ice margin. Leverett (1897) discovered many of these features, and we now recognize more than 70 diagonal changes in streams along the glacial borders across central-southwestern Ohio.

The oldest, highest, and broadest of the criss-crossing abandoned valleys is that called the Teays (detailed description in Tight, 1903) after the present river of that name in West Virginia.

Goldthwait, R. P., 1991, The Teays Valley problem; A historical perspective, *in* Melhorn, W. N., and Kempton, J. P., eds., Geology and hydrogeology of the Teays-Mahomet Bedrock Valley System: Boulder, Colorado, Geological Society of America Special Paper 258.

Only 30 percent of the exposed, but abandoned, Teays trunk valley through West Virginia and southern Ohio is in fact unoccupied by any main stream today. But the breadth (1.5 to 2.5 mi), depth below hilltops of 200 to 300 ft, and vein-like branching tributaries make the old course abundantly clear. There are even potholes at one place under covering sediment. The magnitude and geometry of this old valley indicate it was a main continental drainage developed during many millions of years (Miocene and Pliocene Epochs or longer) (Fig. 1).

Where the rock floor (strath) of this old fossil valley was exposed by the cutting of subsequent crossing river systems (e.g., just east of Piketon, Ohio, today), it was noted that the slope of the strath declined toward the northwest. Wells drilled in an attempt to reach water near this strath floor showed a gradual decline of 0.75-ft/mi northwestward (general elevations in Fig. 1). By the time the detailed work of Stout and others (1943) was summarized, all agreed that the Teays River and tributaries had flowed northwestward in 180° opposition to the present

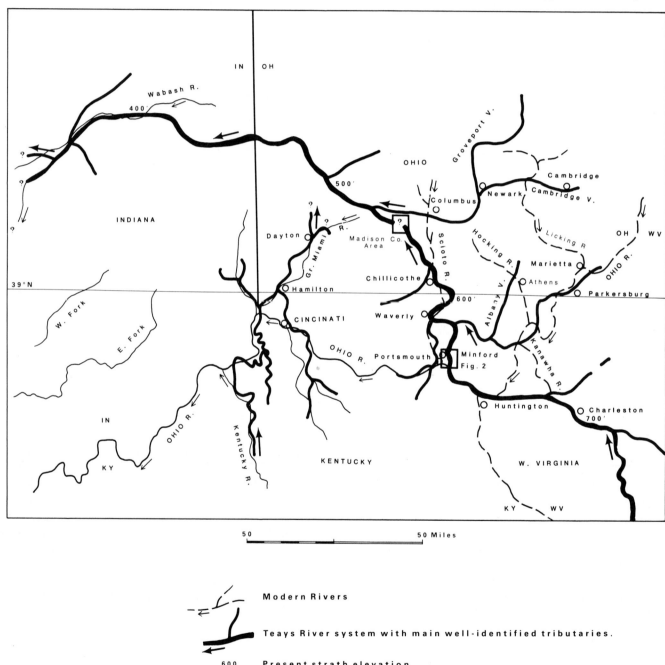

Figure 1. Map of the main Teays System where strath is broad and incontrovertably traced from West Virginia south of Charleston at 740 ft above sea level to west-central Indiana where it is 350 ft above sea level today. Only long tributaries are shown; no detailed headwaters.

lower Scioto River, and at more than a 90° divergence across other main drainages.

COURSE BENEATH THE GLACIAL DRIFT

In the gradual descent northwestward, this great fossil valley abruptly disappears at Chillicothe, Ohio. Farther northward it is completely underground, invisible beneath Wisconsin glacial drift (15,000 to 23,000 yr old). Southeastward, for 10 mi to Londonderry-Richmondale and Waverly, Ohio, the old valley is partly filled with older (probably >125,000-yr-old) Illinoian glacial drift. Even where exposed, half the depth of the bedrock Teays Valley is filled by fluvial, lacustrine, and mass wastage sediments. It is inconceivable that this giant river simply just stopped at Chillicothe; for most of a century, geologists have been arguing the question "Just where did the river go?"

In Ohio, at least, the matter has been considered pretty well settled. As records of farm water wells increased by the thousands, county water reports and geological reports traced the old valley northward through Ross and Pickaway Counties, and northwest through Clark and Champaign Counties, tentatively at first (Stout and others, 1943; Ver Steeg, 1946), then in inescapable detail from 1948 to 1985. Still, there remained a "break" in Madison County. Many followed Coffey (1958), who believed the Teays actually headed due north through Franklin County to a rock-cut trench beneath some northward-flowing river, perhaps the Cuyahoga River flowing into the Lake Erie Basin. But no deeply buried pass or "col" northward has been defined, whereas geophysical resistivity through Madison County (Norris and Spicer, 1958) has identified a 400-ft deep westward-trending trench, though this is too narrow to be defined by existing wells. A second gap, tentatively defined by a very few deep water and oil wells, earlier had demonstrated the presence of an out-of-sight valley of proper depth beneath Shelby, Auglaize, and Mercer Counties (Bownocker, 1899). Thus it was concluded that the Teays River once delivered water to near Decatur in Adams County, Indiana, and continued across Indiana as a major east-west buried valley, forming an S curve (Fidlar, 1943). Some refinement was outlined by Wayne (1956), published by Gray (1982), and simplified herein in Figure 1. Tracing problems do not end in western Indiana, however, for under Lafayette in Tippecanoe County there is a major, broad area of bedrock strath lying as low as 400 ft altitude. This strath may extend farther west into Illinois and join the large Mahomet Buried Valley, as projected by Horberg (1945). The latter valley averages 6 mi in width, 200 to 300 ft deep, and is cut in bedrock nearly all the way. Thus, in the very center of Illinois, the Teays met the Ancient Mississippi River. So goes the traditional interpretation of the Teays System.

WHY DRAINAGE REARRANGEMENT?

With an east-west course near latitude 41°N for 160 mi, the Teays Valley was very vulnerable to glaciation in the Pleistocene Epoch. At least three or four ice advances extended as far as 39°N. One can hardly argue that glaciation did not interrupt Teays River flow! At least it would have interrupted any riverine occupation of the Teays Valley. A few early studies would have us believe that the Teays System terminated prior to glaciations anyway, owing to piracy resulting from regional tilting of the land southwestward. However, recent studies (Hoyer, 1976) indicate this cannot be, for the lacustrine clays called "Minford," which half filled the exposed valley in southern Ohio, have generous proportions of minerals common to three or four glacial tills, as well as much sericite derived from local rock clays. Furthermore, the clays grade into silts, and thence into fine-grained sand as lacustrine deposits are traced northwestward under the glacial tills of west-central Ohio. In some deep test drilling in a tributary entering the Teays under Dayton, Ohio (Spieker, 1961), in the main trench north of Springfield, Ohio, and in the trench across Indiana, material beneath the tills has the variety of rock-mineral composition and sand-pebble grain sizes characteristic of glacial outwash. Some early glacier assuredly loaded these trenches with glacio-fluvial sediments as it reached the valley. Several glaciations collectively have obscured completely all valley topography north of 39°N, yet present streams flow southward, consequent upon the surface topography of the most recent glacial drift.

What, then caused a complete rearrangement of stream courses in unglaciated southern Ohio? The new "Deep Stage" streams passed directly through bedrock hills as much as 50 to 300 ft above the modern valleys (Fig. 2). Now, Deep Stage is the deepest downcutting by streams in mid-Pleistocene time; where drift is thin or absent, this established the modern stream routing and channels cut as much as 100 ft below the Teays strath just south of Chillicothe, Ohio. The classical answer was stream capture, generated by the new, shorter westward route of the Ohio River to the Mississippi Embayment (Fenneman, 1938). A more immediate and instantaneous answer is an older, deeper filling of the valley with glaciolacustrine sediments, upon which new consequent streams developed as governed by lake-bottom topography. This idea was discussed with me by John L. Rich of the University of Cincinnati, who died before publishing his manuscript; but how else could small streams between Wheelersburg and Minford, to Beaver and Piketon, Ohio (Fig. 2), have developed from narrow defiles between high bedrock hills across the broad, soft-sediment-laden Teays Valley? Slow capture would have favored development of medium-sized, broad tributaries up and down the length of the fossil valley. This is not the case; they "dive" in and out of the hills without respect to already existing valleys.

The surface level of lacustrine fill does not cross these divides today, but does change irregularly, both across and up and down the Teays Valley, and reaches the 810-ft contour level in a few tributary valleys (Stout and Schaaf, 1931). A necessary corollary, if this hypothesis be true, is that the original fill extended to somewhere between the 900- and 1,000-ft elevation of today, and has been eroded irregularly downward as much as 200 to 300 ft. Thus was post-Lake Tight drainage consequent on a slightly

AMOUNT AND TIME OF LACUSTRINE FILLING

By 1900, Tight (1903) discovered, through detailed studies in sorthern Ohio and the Huntington-to-Charleston, West Virginia, section of the Teays Valley, that indeed there was once a huge lake; it was subsequently called Lake Tight in his honor. Janssen (1953) reinvestigated the lake bed, and concluded that it ended at 840 ft elevation on the New River section below Hawk's Nest, West Virginia. Thus the lake in the trunk stem was more than 250 mi long. Janssen concluded "the surface of the water . . . may have been appreciably greater." Some laminated, clay-silt fill showed also that the lake extended up all the major exposed tributaries, especially the "fossil" Marietta and Albany

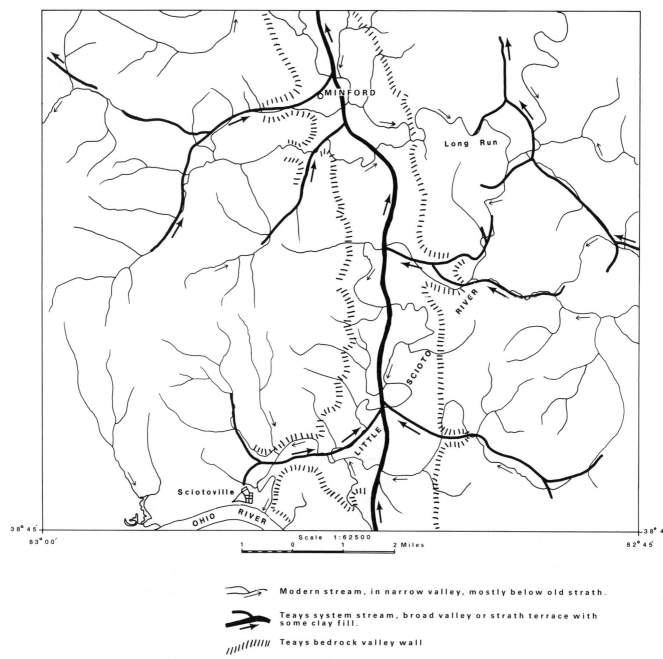

Figure 2. Map of 10 mi of the main Teays Valley northeast of Portsmouth, Ohio. This is a hanging valley where the strath slopes northward from present Ohio River. It is filled with 50 to 125 ft of fluvial and lacustrine sediments (type locality Minford Silt) covered with 1 to 6 ft of loess, or colluvium near rock walls.

rivers. These major channels, together with all irregular, lesser valleys containing patches of clay (Stout and Schaaf, 1931) yielded a several thousand-mile-long, intricate, finger-like shoreline (Wolfe, 1942). The maximum number of couplets (silty clay to fine clay) is possibly 23,000 as projected by Janssen (1953), or more likely 4,000 as actually measured by Hoyer (1972), for these couplets are not necessarily varve years. If half the lacustrine varves (couplets) have been removed by erosion, we surmise that Lake Tight may have endured for 10,000 to 20,000 yr.

When did Lake Tight exist? Not during the last or Wisconsinan glacial invasion, which is well dated by radiocarbon near Chillicothe as only 17,000 to 23,000 yr old (Dreimanis and Goldthwait, 1973; Quinn and Goldthwait, 1985). These Wisconsinan deposits completely obscure the northern portions of the valley. The penultimate glaciation, just prior to 125,000 yr ago, and identified by soils as Illinoian, extended even farther south than the Wisconsinan. It "nosed" down the Teays Valley southeast of Chillicothe, and backed slackwater sediments into the older, deeply eroded Lake Tight sediments, where they were largely eroded away (Rhodehamel and Carlston, 1963). One yet older glaciation, long-called Kansan and probably more than 700,000 yr old, is identified in the vicinity of and southward from Cincinnati; the deposits rest on Teays System slopes, and appear to postdate "The Old Kentucky" tributary (see Teller and Goldthwait, this volume). Across Indiana, classical "Kansan-age" drift has been identified in borings in what has been asserted to be the buried Teays Valley. The Mahomet Valley definitely shows four glacial stages of filling by outwash and tills (Horberg, 1950).

The clincher is the recent discovery of reversed polarity orientation of magnetic particles in unaltered Lake Tight clays. This was first measured by Hoyer (1972) near the type "Minford Silt" locality, and properly confirmed east of Huntington, West Virginia, by Bonnett and others (1978). The latter evidence is presented in detail in this volume. It is safe to conclude that flooding, filling, and drainage reversal of the exposed portion of the upstream part of the Teays drainage system occurred more than 700,000 yr ago by action of an early Pleistocene glacier. This event is construed herein to have occurred rather rapidly by superposition of lake floor sediments. Subsequently, lacustrine sediments were erosionally reduced by nearly half their volume. There are other hypotheses which only time and further evidence will clarify.

DISCOVERING MAJOR TRIBUTARIES

Ever since the major trunk valley was defined, geologists have been finding and arguing about various abandoned tributary streams. These cut across the present drainage divides and, where exposed, also contain lacustrine Minford clays and silts. The fertile farm lands developed on these lacustrine deposits in otherwise unglaciated areas are deforested and can be traced readily among sandstone or shale hills well above the level of the present floodplains. One of the longer tributaries, the "Marietta River" (Fig. 1) headed more than 100 mi farther east, above Parkersburg, West Virginia. It is followed in part by OH Route 7 to Pomeroy, Ohio, then by the Ohio River itself to Gallipolis, and then by US Route 35 to Jackson and OH Route 124 to Beaver (detail in Stout and others, 1943). A major northeast branch, "Albany River," started northwest of Athens, Ohio, and in general is followed by US 50 and OH Routes 689-160-329 to Rio Grande, Ohio. Obviously these dendritic valleys, now high and dry, made ideal road beds that rest on an undulating surface of clay–silt–fine sand fill.

Almost as obvious is the hypothesis that the Ohio River, and other streams, were superimposed across divides upon portions of Teays trunk and tributaries.

Even longer, exposed southern tributaries such as the "Old Kentucky River" were ignored by most early 20th century geologists because it was thought that if, indeed, they joined the Teays Valley, the juncture was under glacial deposits farther north. If so, this river ran northward up what today is Great Miami River Valley, passing under drift from Middletown to Germantown (OH Route 4), and around to Miamisburg, Ohio, where there is just the right strath level (450 to 500 ft) to carry the "Old Kentucky." A short, mile-wide tributary at the appropriate juncture angle joins under Fairborn, Ohio. But water resources studies (Norris and Goldthwait, 1948; Goldthwait, 1950) found no corridor north and northeastward to join the Teays trunk river. Modern geophysical exploration now is being used in an attempt to trace this problem valley. All these studies do show a strath (bedrock) floor nicely graded to the main Teays Valley, and help confirm the identity and reality of a former major continental river system.

Other tributaries, large and small, are fully covered by glacial drift throughout their lengths. "Groveport River" in central Ohio is one of the longest, almost 150 mi. Starting at Wooster, Ohio, this valley curves southwestward under Loudonville, Fredricktown, and Mt. Vernon, past Newark, and then westward from Canal Winchester to Grove City. By happenstance, routes US 40 and I-70 follow this lower valley to where it joins the buried Teays Valley near London, Ohio. Position of the upper reaches of Groveport River is made clear near the Wisconsinan glacial limit by sandstone hills on either valley side, but the lower westward reach has no surficial expression, and much of it is altered, below cover, by the yet deeper, narrower "Deep Stage" valley, which took a similar course past Newark but continued southward ("Newark River," the ancestral Scioto River). Once bedrock contour maps were completed, untangling of mysteries beneath glacial drift is achieved by knowing valley bottom (strath) gradients and by the angle of stream pattern junctions.

A final puzzle, treated only generally until now, is: Why are tributary straths through Indiana so short, whereas those in Ohio, Kentucky, and West Virginia are so long? The drainage area of the Teays System southeast from Indiana now embraces 87,000 mi^2. Why so narrow and limited to the west? Undoubtedly some tilting has taken place in 10 m.y.: how much, and when?

If glacial drift did not now fill it, the Teays Valley northwest of Chillicothe to London and St. Paris, Ohio, would stand out as the most remarkable 500-ft-deep canyon in Ohio and be the site of numerous state parks.

We now are ready to look at some new details and interpretations revealed and suggested by recent work. Some of the past questions may be answered, some left unresolved, and new questions will be raised.

REFERENCES CITED

Bonnett, R. B., Sanderson, D. D., and Noltimier, H. C., 1978, Paleomagnetic results from the Minford Silt, Teays Depot, West Virginia: EOS Transactions of the American Geophysical Union, v. 59, p. 225.

Bownocker, J. A., 1899, A deep preglacial channel in western Ohio and eastern Indiana: American Geologist, v. 23, p. 178–182.

Coffey, G. N., 1958, Major glacial drainage changes in Ohio: Ohio Journal of Science, v. 58, no. 1, p. 43–49.

Drake, D., 1825, Geological account of the valley of the Ohio: American Philosophical Society Transactions, n.s., v. 2, p. 124–139.

Dreimanis, A., and Goldthwait, R. P., 1973, Wisconsin glaciation in the Huron, Erie, and Ontario lobes, in Black, R. F., Goldthwait, R. P., and Willman, H. B., eds., The Wisconsinan Stage: Geological Society of America Memoir 136, p. 71–106.

Fenneman, N. M., 1938, Physiography of the eastern United States: New York, McGraw-Hill Book Co., 534 p.

Fidlar, M. M., 1943, The preglacial Teays Valley in Indiana: Journal of Geology, v. 51, p. 411–418.

Goldthwait, R. P., 1950, Wisconsin glacial deposits, in Norris, S. E., The water resources of Greene County, Ohio: Ohio Department of Natural Resources, Division of Water Bulletin 19, p. 13–19.

Gray, H. H., 1982, Map of Indiana showing topography of the bedrock surface: Indiana Geological Survey Miscellaneous Map 36, 1 sheet, Scale?

Hildreth, S. P., 1838, in Mather, W. W., 1st Annual Report: Geological Survey of the State of Ohio, p. 50.

Horberg, C. L., 1945, A major buried valley in east-central Illinois and its regional relationships: Journal of Geology, v. 53, p. 349–359.

——— , 1950, Bedrock topography of Illinois: Illinois State Geological Survey Bulletin 73, 111 p.

Hoyer, M. C., 1972, Remanent magnetism of Minford Silt, southern Ohio: Geological Society of America Abstracts with Programs, v. 4, p. 554.

——— , 1976, Quaternary valley fill of the abandoned Teays drainage system in southern Ohio [Ph.D. thesis]: Columbus, Ohio State University, 146 p.

Janssen, R. E., 1953, Varved clays in the Teays Valley: West Virginia Academy of Science Proceedings, v. 25, p. 53–54.

Leverett, F., 1897, Changes in drainage in southern Ohio: Denison University Scientific Laboratory Bulletin 9, part 2, p. 18–21.

Norris, S. E., and Goldthwait, R. P., 1948, The water resources of Montgomery County, Ohio, with a section on glacial geology by R. P. Goldthwait: Ohio Water Resources Board Bulletin 12, 83 p.

Norris, S. E., and Spicer, H. C., 1958, Geological and geophysical study of the preglacial Teays Valley in west-central Ohio: U.S. Geological Survey Water Supply Paper 1460-E, p. 199–232.

Quinn, M. J., and Goldthwait, R. P., 1985, Glacial geology of Ross County, Ohio: Ohio Geological Survey Report of Investigations 127, 42 p.

Rhodehamel, E. C., and Carlston, C. W., 1963, Geological history of the Teays Valley in West Virginia: Geological Society of America Bulletin, v. 74, p. 251–273.

Spieker, A. M., 1961, A guide to the hydrogeology of the Mill Creek and Miami River Valleys, Ohio, in Guidebook for field trips, Cincinnati meeting of the Geological Society of America, p. 217–237.

Stout, W. E., and Schaaf, S. D., 1931, The Minford Silts of southern Ohio: Geological Society of America Bulletin, v. 42, p. 663–672.

Stout, W. E., Ver Steeg, K., and Lamb, G. F., 1943, Geology of water in Ohio: Ohio Geological Survey, 4th Series, Bulletin 44, 694 p.

Tight, W. G., 1903, Drainage modifications in southeastern Ohio and adjacent parts of West Virginia and Kentucky: U.S. Geological Survey Professional Paper 12, 111 p.

Ver Steeg, K., 1946, The Teays River: Ohio Journal of Science, v. 46, no. 6, p. 297–307.

Wayne, W. J., 1956, Thickness of drift and bedrock physiography of Indiana north of the Wisconsin glacial boundary: Indiana Geological Survey Report of Progress 7, 70 p.

Wolfe, J. N., 1942, Map compiled from 900 foot contour on 50 U.S. Geological Survey Quadrangles showing "one possible stage of Lake Tight": Ohio Journal of Science, v. 42, no. 1, p. 2–12.

MANUSCRIPT RECEIVED BY THE SOCIETY JUNE 29, 1990

A paleomagnetic study of the early Pleistocene Minford Silt Member, Teays Formation, West Virginia

Richard B. Bonnett
Department of Geology, Marshall University, Huntington, West Virginia 25755
Hallan C. Noltimier
Department of Geological Sciences, Ohio State University, Columbus, Ohio 43210
Dewey D. Sanderson
Department of Geology, Marshall University, Huntington, West Virginia 25755

ABSTRACT

At some time during the Pleistocene Epoch, a part of the modern Ohio River drainage system in Ohio and West Virginia developed in response to impoundment of the ancestral Teays River drainage system. Rhythmites formed in the lacustrine slackwaters and remain today, extending as much as 150 to 200 km upstream from the Pleistocene ice front, in Teays Valley, West Virginia. A total of 303 oriented paleomagnetic specimens represent a composite stratigraphic section from the Minford Silt Member of the Teays Formation in Teays Valley. Of these, 224 specimens carry a stable reversed magnetization due to detrital magnetite and hematite. Two distinctive lithologic intervals with definitive magnetic intensities were found in the stratigraphic section; thick, light-colored rhythmites carry six times more remanent magnetization intensity than thin, dark-colored rhythmites, reflecting variations in the ratio of magnetite to hematite. From the Pleistocene paleomagnetic chronology, the glacial diversion of the Teays River in Ohio and West Virginia took place during the Matuyama reversed polarity chron, in the time interval between 0.79 and 1.60 Ma, the change attributed to an Early Pleistocene age, most probably the F or G glaciation. We propose that the Minford Silt was deposited between 0.79 and 0.88 Ma.

INTRODUCTION

In pre-Pleistocene time, a major river system and its tributaries, named the Teays, drained large areas of the states of Virginia, North Carolina, West Virginia, Kentucky, Ohio, Indiana, and Illinois (Tight, 1903). The river was thought to have flowed generally northwestward, joining the preglacial Mississippi in west-central Illinois about 40 miles north of Springfield (Thornbury, 1965; Kempton and others, this volume). During the Pleistocene, one of the early glacial advances impounded the Teays drainage system in south-central Ohio, near Chillicothe. The blocking of the river formed Lake Tight in Ohio, which extended southeastward to approximately Hawk's Nest, West Virginia. In West Virginia, this lake is frequently called Lake Teays.

The sediments of this fluvial and lacustrine system were first named the Teay Formation by Campbell (1900), the type locality being Teays Depot, West Virginia (Fig. 1). Today the term Teay is considered archaic; the preferred name is Teays. The lower Gallia Sand Member and the upper Minford Silt Member were differentiated in Ohio by Stout and Schaaf (1931). The stratigraphic equivalence of the units from Ohio to West Virginia was established by Janssen and McCoy (1953), and by Rhodehamel and Carlston (1963).

The Minford Silt, which represents the slackwater lacustrine phase formed by impoundment of the former Teays River, has been found to be mainly clay-size material at the type locality. Upstream from the point of blockage, a cold periglacial climate likely prevented lacustrine seasonal overturn and permitted the

Bonnett, R. B., Noltimier, H. C., and Sanderson, D. D., 1991, A paleomagnetic study of the early Pleistocene Minford Silt Member, Teays Formation, West Virginia, *in* Melhorn, W. N., and Kempton, J. P., eds., Geology and hydrogeology of the Teays-Mahomet Bedrock Valley System: Boulder, Colorado, Geological Society of America Special Paper 258.

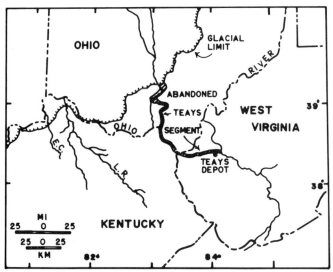

Figure 1. Index map showing site location at Teays Depot, West Virginia, and abandoned segments of the preglacial Teays River in Ohio and West Virginia. E.C. = Eagle Creek; L.R. = Licking River.

deposition of varved or rhythmitic fine-grained sediments, or rhythmites. Ultimately, the impounded water was pirated, and then, through a complex sequence of events, the present Ohio River system evolved. If the rhythmites are varves, their count would be an estimate of the time during which the river was impounded. The stratigraphic section of this study yields an estimate of 7,000 yr, which agrees favorably with Janssen and McCoy's (1953) estimate of 6,500 yr. In several segments of the stratigraphic section, the rhythmites display an increase in thickness with a cyclicity of every 9 to 11 couplets, suggesting they are varves. This could represent climatic influence on precipitation and, hence, sedimentation rates by sunspot activity.

The preglacial Old Kentucky River was the major southern tributary of the Teays River in northern Kentucky and southwestern Ohio (Teller, 1973; Teller and Goldthwait, this volume). Two ancestral tributaries of the Old Kentucky River, Eagle Creek and Licking River (Fig. 1), reveal lacustrine clays that correlate with Teays-age deposits. In turn, a deposit overlying the clay has tentatively been identified as Kansan (pre-Illinoian) till (Durrell, 1961).

Some pre-Illinoian till is found south of the Ohio River in northern Kentucky (Leverett, 1902; Durrell, 1961; Ray, 1963, 1969; Teller, 1973). Illinoian till has been mapped in the Ohio River Valley (Goldthwait and others, 1961). One or more early pre-Illinoian glaciers advanced far enough to block and divert the Teays River and its tributaries. Goldthwait and others (1961) and Flint (1971) favor a "Kansan age" (second glaciation), whereas Leighton and Ray (1965) and Swadley (1971) favor a "Nebraskan age" (first glaciation). Richmond and Fullerton (1986) recommend abandonment of the terms "Nebraskan" and "Kansan" because of multiple correlation problems. It has generally been agreed that breaching of divides drained the various lakes and established the Ohio River drainage system.

Stratigraphy and clay mineralogy studies by Ettensohn (1974) and Ettensohn and Glass (1978) clearly indicate that the impoundment of the Teays tributaries resulted in deposition of bedded clay and silt, which were later covered by pre-Illinoian tills. The area of this study is along the main trunk of the old Teays River, which would have been subjected to the same type of slackwater deposition as occurred in the Cincinnati region. However, our study area is no closer than 125 km to the nearest Pleistocene ice margin and, consequently, has no overlying or interbedded tills that help to date the clays.

SITE LOCATION AND DESCRIPTION

The study area is in an abandoned segment of the Teays River Valley at Teays Depot (38.44°N, 81.95°W), West Virginia. Two sites within 300 m of each other make up the sampled stratigraphic section, locally called the Teays Clay (equivalent to the Minford Silt of Stout and Schaaf, 1931). Due to differences in site elevations, an unavoidable interval of 3 m was not sampled. The rhythmites were sampled from their base where they unconformably overlie the Gallia Sand member (elevation, 214 m) to the top of their outcropping, a stratigraphic interval of 19 m.

In West Virginia, the highest known elevation for exposed Minford Silt is in the Rock Step Run valley, approximately 8 km east-northeast of Teays Depot, at an elevation of 247 m (Bonnett, 1975). The Rock Step Run site is on the north valley side, and the Teays Depot site is on the south. Attempts to identify beach zones that would delineate upper lake limits, and thus establish water depth, have not been successful.

The section sampled is 19 m thick and is believed to represent most, but clearly not all, of the sediments deposited during impoundment. Rhodehamel and Carlston (1963) reported a maximum section thickness of slightly more than 36 m and a maximum continuous exposed thickness of slightly more than 31 m at Mt. Vernon, 1.5 km northeast of Teays Depot. Near the top of our section there is a thickening of the rhythmites, the presence of several conformable and interbedded silt layers, and a capping of loess. Clearly, a change of depositional environments is indicated by the upper part of the section, suggesting a late lacustrine to eolian transition and eventual abandonment of the lake. How much erosion of the clay has taken place is highly conjectural; however, the site at Teays Depot is in the eastern one-third of Teays Valley where preservation of the Teays Formation is most complete. Loess has not generally been recognized this far south (Thorp and others, 1952); however, not far downstream in the abandoned Teays River valley between Waverly and Minford, Ohio, loess has been reported (Bigham and others, this volume).

SAMPLE COLLECTION

A special aluminum jig was constructed to retrieve a series of eight samples per setup (Fig. 2). Vertically cut faces perpendicular to bedding were prepared for the jig, which was

Figure 2. Aluminum sampling jig constructed for collection of eight varved sediment specimens (2.54 cm diameter, 2.54 cm length) at 5.0 cm intervals.

TABLE 1. PILOT GROUP PALEOMAGNETIC RESULTS BY POLARITY AT VARIOUS AF DEMAGNETIZATION STAGES

AF	DEC	INC	κ	α_{95}	J/J_0	BSI
Group a: Reversed Polarity (N = 21)						
NRM	194.7	-40.7	4.1	18.1	1.0	
						0.76
100	188.9	-35.2	8.5	11.6	1.1	
						0.84
200	188.6	-37.7	13.9	8.9	1.1	
						0.91
300	187.7	-38.0	12.2	9.5	1.0	
						0.92
400	187.2	-39.4	13.8	8.9	0.97	
						0.92
500	187.4	-40.5	13.1	9.1	0.93	
						0.94
500	187.0	-40.6	13.7	8.9	0.89	
						0.90
700	187.0	-37.9	15.0	8.5	0.93	
Group b: Normal Polarity (N = 11)						
NRM	338.6	61.3	10.3	14.9	1.0	
						-0.35
100	353.1	63.4	28.2	8.8	2.1	
						0.86
200	350.7	63.3	21.1	10.2	1.8	
						0.90
300	353.0	61.7	24.2	9.5	1.7	
						0.87
400	346.2	60.3	21.6	10.1	1.6	
						0.85
500	341.4	59.4	16.4	11.6	1.4	
						0.76
600	324.8	58.5	5.1	22.3	1.4	
						0.63
700	308.8	51.5	3.0	32.1	1.5	

affixed to the outcrop by aluminum spikes. Cylindrical Lexan plastic sleeves, with an outer diameter of 2.5 cm and a length of 2.5 cm, were guided into the moist clay parallel to bedding. Sample orientations were determined with a Brunton compass. The samples were then sealed with a paraffin film covering. A total of 339 samples was collected, but due to some overlap of the sampling jig positions between prepared outcrop faces, 303 samples represent the entire stratigraphic section analyzed in this study.

Each cylindrical sample contained an average of six to eight rhythmites; samples were collected at intervals of 5 cm. Each sample represents a weighted average since rhythmites near the cylindrical axis of the specimen are represented by a greater width and hence a larger volume. Couplet thicknesses range from 1.5 to 15.0 mm.

PALEOMAGNETIC PROCEDURES

The remanent magnetization of the samples was measured with both cryogenic and spinner magnetometers in the Paleomagnetism Laboratory at The Ohio State University. Thirty-two samples (ca. 10 percent) were initially selected from the sampled section for preliminary detailed measurement and study of natural remanent magnetization (NRM) and response to alternating field (AF) treatment in 100-gauss (10 mT) increments to 700, 850, and 1,000 gauss (Bonnett and others, 1978, 1983). The results of the preliminary analyses formed the basis for the completed study. Thermal demagnetization methods were not attempted because we did not want to destroy the integrity of the sealed samples nor generate an artificial chemical remanent magnetization (CRM) by dehydration of limonite and goethite likely to be present in the samples.

RESULTS

The NRM and AF results for the initial 32 samples are given in Table 1. The AF results are shown through 700 gauss and were separated into a normal polarity subgroup of 11 samples, and a reversed polarity subgroup of 21 samples. The data from the two subgroups were plotted on a pair of equal area nets, one for each AF treatment, and the J/J_0 (residual/original remanence) ratios were plotted after each AF treatment. A Zijderveld diagram was constructed, and the mean Briden Stability Index (BSI) was calculated for each group at each 100-gauss AF increment (Zijderveld, 1967; Briden, 1972). These data were used to choose the best interval of AF treatment for the Minford Silt samples. The results between 200 and 600 gauss exhibit the most consistent overall characteristics for stable remanence and reliable mean direction. Figure 3a shows the 300-gauss AF data net for the reversed polarity group; Figure 3b, the 400-gauss AF data for the normal polarity group. Figure 4a and 4b show the Zijderveld plots for the normal and reversed groups. Figure 5 shows the

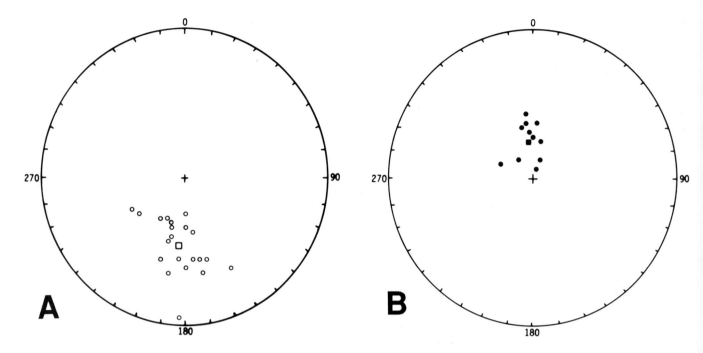

Figure 3. A, equal area net plot of the test group results for the reversed-polarity subgroup at 300 Oe. peak AF. The open square represents the direction of the reversed-polarity axially symmetric geomagnetic field at the sampling locality. B, equal area net plot of the test group results for the normal-polarity subgroup at 400 Oe. peak AF. The closed square represents the same geomagnetic field with normal polarity.

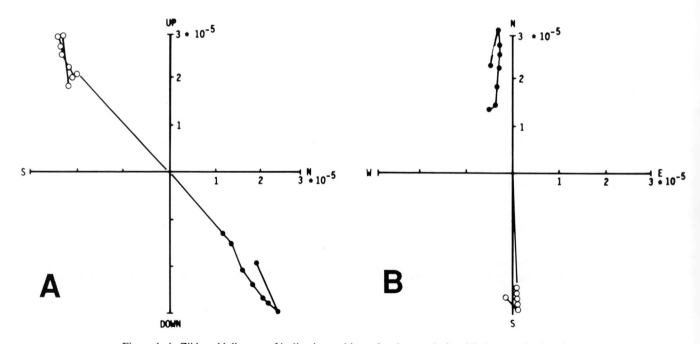

Figure 4. A, Zijderveld diagram of inclination and intensity changes during AF demagnetization, from the NRM through 700 Oe. peak, for the reversed (open circles) and normal (solid dots) subgroups. B, Zijderveld diagram of declination and intensity changes, from NRM through 700 Oe. peak, for the reversed (open circles) and normal (solid dots) subgroups.

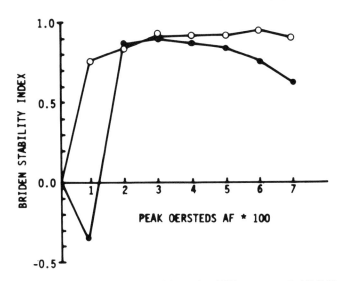

Figure 5. Plot of the Briden Stability Index (BSI) versus peak AF field (NRM through 700 Oe.) for the normal polarity (solid dots) and reversed polarity (open circles) subgroups.

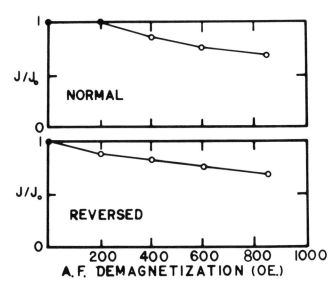

Figure 6. Relative decay of magnetization (J/J_o) as a function of AF demagnetization treatment.

BSI values plotted versus applied field. Table 1 summarizes the data plotted in Figures 3, 4, and 5. The results from the 10 percent test group indicate that both polarities are present, which is highly unlikely under normal geologic conditions for the 7,000-yr depositional record represented by the clays.

The remaining samples were AF demagnetized at 200, 400, 600, and 850 gauss, and the entire suite of 303 samples, from NRM through 850 gauss, is given in Table 2 in terms of a 79-sample normal, and a 224-sample reversed subgroup.

EVALUATION OF THE PALEOMAGNETIC DATA

Seventy-nine of the entire sample suite have positive inclinations (downward to the north), whereas 224 samples have negative inclinations (upward to the south). The normal polarity subgroup has an α_{95} of 14°, whereas the reversed polarity subgroup has an α_{95} of 4°. The smaller α scatter in the reversed group is due in part to a larger sample size (Fisher, 1953).

Because the span of time represented by the sampled section is brief, it is unlikely that the section records a polarity change. This, plus the statistical results, leads us to the interpretation that the main group with reversed polarity represents the primary depositional remanent magnetization (DRM) of the sediments, and the smaller normal polarity group represents a CRM acquired postdepositionally in the zone of weathering.

During AF demagnetization, a greater stability of remanent magnetization is shown by the reversed subgroup. This is indicated by several lines of evidence. (1) The BSI values are consistently greater for the reversed subgroup (Table 1, Fig. 5). (2) The values of α_{95} are uniformly smaller for the reversed subgroup, both in the test group (Table 1) and in the complete result subgroup (Table 2). This comparison is partially biased because of the difference in sample numbers between groups, but the values of κ, the invariance, are more uniform in the reversed test group (Table 1) and are uniformly larger in the complete reversed group (Table 2). The complete normal subgroup residual magnetization has a standard deviation about 25 percent larger than the complete reversed subgroup, a fact consistent with the BSI results for the smaller test subgroups. The reversed test subgroup has a BSI more than 0.90 through 700 gauss which simply states that the reversed remanent direction and intensity change by less than 10 percent during AF demagnetization.

The relative decay of remanent intensity is plotted versus AF treatment in Figure 6 for both normal and reversed subgroups. In this comparison, both show remarkable stability; the remanence decreases by only 30 percent at 850 gauss. The reversed subgroup has slightly lower J/J_o ratios after 200 and 400 gauss, but the general nature of both subgroup's response to AF treatment suggests similar magnetic mineralogy. Figure 7 shows the behavior of κ and α_{95} with AF treatment for both polarity groups.

The mean paleomagnetic inclination of the reversed group, at 850 gauss, is 40°. The calculated inclination from the axial dipole formula is 60°, and the present nonaxial field inclination is 71°. The reversed suite of specimens appears to show the common inclination flattening of DRM (Irving, 1967) due to postdepositional compaction. The normal polarity group has an inclination of approximately 73°, just slightly off the present magnetic field inclination, and consistent with late Pleistocene or Holocene acquisition of a CRM.

Saturation isothermal remanent magnetization (SIRM) analysis was carried out on specially prepared samples taken from the Teays section and consisting of thicker and thinner laminae. In this analysis, the paleomagnetic samples were subjected to increasing direct current (DC) magnetic fields in equal increments at room temperature, and the resulting increase in isothermal

TABLE 2. SUMMARY OF PALEOMAGNETIC RESULTS BY POLARITY GROUP AT VARIOUS STAGES OF A.F. DEMAGNETIZATION

AF	DEC	INC	$J*10^{-5}$ (gauss cm^3)	J/J_o	κ	$\alpha 95$	VGP (lat.)	VGP (long.)
Group a: Reversed Polarity (N = 224)								
NRM	193.4	−33.5	2.79	1.00	5.57	4.40	−66.8	−115.9
200	192.5	−37.8	2.46	0.88	6.78	3.95	−69.7	−117.5
400	192.9	−39.3	2.26	0.81	6.53	4.03	−70.4	−120.1
600	192.7	−39.4	2.12	0.76	8.27	3.52	−70.6	−119.7
850	193.5	−40.1	1.94	0.70	8.22	3.50	−70.6	−122.4
Group b: Normal Polarity (N = 79)								
NRM	289.0	73.7	2.60	1.00	2.95	11.77	41.7	−121.6
200	304.1	75.3	2.58	0.99	2.64	12.76	49.0	−117.8
400	304.0	72.4	2.24	0.86	2.46	13.32	49.4	−125.0
600	292.1	73.1	1.97	0.76	2.34	14.00	43.2	−123.3
850	284.2	74.3	1.85	0.71	1.75	18.37	39.5	−120.0

remanent magnetization is plotted versus applied field strength. The results can often determine the magnetic mineralogy, or place useful limits on the range of possibilities (Bloemendal and others, 1979; Walling and others, 1979).

Two groups of paleomagnetic samples were prepared from (A), the thinner, dark-colored varve sequence, and (B), the thicker, light-colored varve sequence, which were taken from an unweathered column of Teays Formation, as shown in Figure 8. In SIRM analysis, the intensity of acquired isothermal remanent magnetization (IRM) is significant, not the direction. The SIRM results from two samples are shown in Figure 9. Both paleomagnetic samples were given an initial AF demagnetization of 100 gauss to remove soft viscous magnetization. They were then each given a unidirectional and isothermal exposure along the cylindrical sample axes to stepwise increasing DC magnetic fields, from 500 gauss to a peak of 3,500 gauss, in increments of 500 gauss for intervals of 1 hr, followed immediately by remanence measurement. The results for the thinner laminated varve sequence are quite distinct from the thicker laminated varve sequence. The thin varves show saturation of the artificial IRM by 2,500 gauss, whereas the thick varves show no tendency to reach saturation of artificial IRM by 3,500 gauss. At 3,500 gauss, the IRM intensity of the thin varve material is one order of magnitude greater than the thick varve material (5 versus 0.5 milligauss). These results clearly indicate that the proportion of magnetic minerals in each varve sequence is different. The thinner varve couplets, which reach saturation by 2,500 gauss, owe much of their remanence to magnetite, whereas the thicker varve couplets owe their remanence mainly to hematite. This is quite independent of NRM results, which show that the thicker varves have about six times the intensity of the thin varves and carry more total iron. This is discussed below, with the clay mineralogy and iron content.

DISCUSSION OF CLAY MINERALOGY AND IRON CONTENT

We believe that the primary remanence of the measured section is a reversed DRM, dating from deposition during an interval of reversed geomagnetic polarity, while the normal polarity subgroup represents postdepositional chemical alteration of the detrital magnetic minerals and acquisition of secondary remanence during the Brunhes normal-polarity chron. Figure 10 is a log of the remanent intensity, declination, and inclination versus stratigraphic position. The mixed polarities are concentrated in the upper portion of each outcrop section. Below the zones of apparent postdepositional alteration, polarities are predominantly reversed. The scattered normal polarities are due to either weathering during the Pleistocene in the proximity of joints and bedding planes, or weathering in Recent or historic times, since the time of exposure of the sediment in the railway cut outcrop, and possibly all three.

The high resistance of these specimens to AF demagnetization is characteristic of remanence caused mainly by fine-grained hematite, which has a high coercive field. We estimate that 75 percent of the DRM is carried by hematite and 25 percent by detrital magnetite (with a much lower coercive field). Clastic hematite is expected in the fine-grained, light colored upper portion of the varve couplet because the color of the dry varves varies from pale red (10R 6/2) to light brownish gray (10YR 6/2) and suggests the presence of hematite, as well as oxyhydroxides of iron such as limonite or goethite. Because the iron oxides are so fine grained and are present in minor amounts, their presence could not be independently confirmed or resolved by optical examination or x-ray diffraction.

The mineralogic composition of the varved sediments has been previously investigated by others. Table 3 gives the

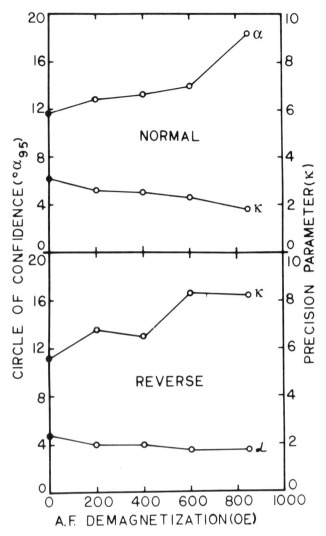

Figure 7. Response of the two magnetic polarity groups to demagnetization as expressed by κ, the precision parameter and α_{95}, the circle of confidence at the 95 percent level.

Figure 8. One of the stratigraphic transitions where there is an abrupt change in color, thickness, and magnetization of the varves.

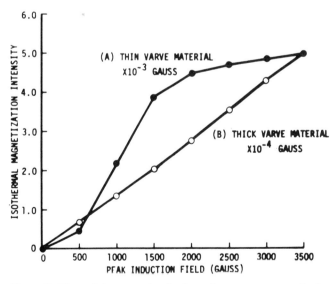

Figure 9. Plots of the saturation isothermal remanent magnetization (SIRM) results for samples of the thick (open circles) and thin (solid dots) varve material shown in Figure 9. The DC magnetic field ranged from 0 to 3,500 gauss, and was applied unidirectionally along the axes of the cylindrical subsamples.

generalized results of Hoyer (1976) from central southeast Ohio. Analyses from southwestern Ohio (Ettensohn and Glass, 1978) reveal less illite and chlorite but a higher content of expandable clays than reported by Hoyer (1976). Determination of the composition of the clay fraction by Rhodehamel and Carlston (1963) near our locality shows 40 percent kaolinite, 30 percent mica (probably muscovite), 20 percent vermiculite, and 10 percent quartz. The only iron-bearing mineral reported by these investigators is chlorite. According to Hoyer (1976), the lower gray clays have an average chlorite content of 13 percent, while the upper brown clays have 6 percent. Hoyer attributed the brown group to weathering of the gray group and found that the silt fraction has an average of 5 percent chlorite. The results are in Table 4. X-ray diffraction results from this study were similar to those of Rhodehamel and Carlston (1963), but we did not detect chlorite

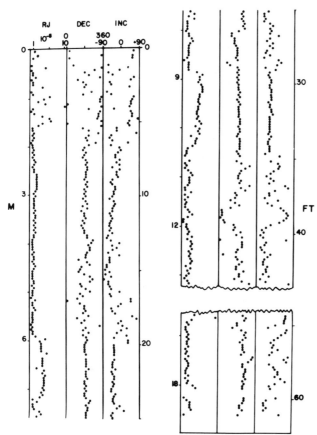

Figure 10. Magnetic stratigraphy, intensity, declination, and inclination of the Minford Silt Member of the Teays Formation. Note: the left column continues downward as the right column, and the irregular break represents the unavoidable 3.0 m of missing section (after Bonnett and others, 1978).

TABLE 3. ANALYTICAL RESULTS OF SIZE AND COMPOSITION OF THE MINFORD SILT FROM CENTRAL SOUTHEASTERN OHIO*

Mineral	Abundance (%)
Clay size (<2 μ)	80 or more
Illite	70–75
Kaolinite	2–11
Chlorite	2–18
Vermiculite	1–12
Smectite	1–3
Quartz	10

*Compiled from Hoyer, 1976.

(see Table 4). The magnetic susceptibility of some randomly selected natural samples from each color group show an average value of 8×10^{-6} cgs units (101×10^{-6} SI units). This value is consistent with a magnetic mineralogy of mainly hematite with lesser amounts of magnetite. Consequently, the combination of a resistance to AF demagnetization, reddish brown pigmentation, and low magnetic susceptibility suggests that overall, hematite is the major carrier of the weak depositional remanent magnetization (DRM).

In the center of the stratigraphic section (Fig. 10), a distinct and regular signature in the magnetic intensity is observed coincident with changes in the varve thickness, as depicted in Table 5. Each of the four transitions in varve thickness and magnetic intensity is abrupt. The thin, darker varves show low remanent intensity, whereas the thick, lighter varves show much higher remanent intensity. Fig. 8 illustrates such a transition.

Magnetic susceptibility measurements of the two groups give similar results. Samples from both thicker and thinner varve couplet intervals were chemically analyzed for total iron content.

The results yielded 5.1 and 4.4 percent total iron content for the thicker and thinner varves, respectively. Total iron content and bulk magnetic susceptibility are proportional to one another. Since the remanent intensity of the thicker and thinner varve sequences differs by a factor of six, the similar bulk iron contents cannot account for the disparity in magnetization. We believe that the explanation for the large difference in the natural remanent magnetization and small difference in iron content is a different magnetic mineralogy. Some iron may be present in the clay mineral fraction and not involved with the remanence.

The SIRM results, discussed previously, show that the thinner varve couplets carry relatively more magnetite than hematite (an order of magnitude greater SIRM intensity at induction of 3,500 gauss). Because magnetite has a much larger specific magnetization than hematite, there might at first appear to be a discrepancy since the thinner varves have less magnetization. The chemical analysis indicates that the thicker varve couplets have more iron. The discrepancy may be explained by a small change in the relative abundance of magnetite. A relative change of 0.7 percent of the iron (5.1 percent versus 4.4 percent) carried in magnetite can account for the six-fold increase in remanent moment in the thick varves. The thin couplets, therefore, carry their magnetization and iron as magnetite, while the thick couplets carry 0.7 percent more magnetite and more total iron as hematite. The SIRM results clearly point to a simple magnetic mineralogy, as proposed above.

APPLICATION TO DRAINAGE OF LAKE TIGHT

The time at which abandonment of former Lake Tight and establishment of the present-day Ohio River System in West Virginia and Ohio took place can be estimated from the Pleistocene paleomagnetic record (Cox, 1969; Cox and others, 1963; Lowrie and Alvarez, 1981) and North American glacial chronology (Richmond and Fullerton, 1986). The paleomagnetic data of our study indicate deposition during reversed geomagnetic polarity, which if it is of Pleistocene age, would be the Matuyama

TABLE 4. X-RAY MINERAL ANALYSES, TEAYS DEPOT*

Mineral	Light Colored Thick Varves (%)	Dark Colored Thin Varves (%)
Kaolinite	41.5	51.5
Chlorite	<1-10†	<1-20†
Illite/mica	33.0	34.0
Montmorillonite	12.5	11.5
Quartz	13.5	2.0
Total	100.5	99.0

*Averages of four determinations.
†Detection threshold: 1 percent for Mg chlorite and 10 percent for Fe chlorite.

TABLE 5. PATTERN IN THE MAGNETIC STRATIGRAPHY IN CENTRAL PORTION OF SECTION*

Varve Thickness (mm)	Color (comparative†)	Time Interval (yr)	NRM Intensity ($\times 10^{-6}$ gauss)
2	Dark	......	7
4	Light	300	42
2	Dark	800	7
4	Light	400	42
2	Dark	......	7

*Entries in table are in stratigraphic sequence.
†Average Munsell Color: dark, 10YR 6/2; light 10R 6/2.

reversed-polarity chron. Geologic evidence suggests pre-Illinoian blockage, which is consistent with the paleomagnetic results. It seems to be generally accepted (Cooke, 1973; Richmond and Fullerton, 1986) that the Illinoian stage commenced during the Brunhes normal-polarity chron and is currently dated at 0.302 Ma (Bowen and others, 1986). Pre-Illinoian glaciation commenced either in the late Gauss normal- or early Matuyama reversed-polarity chrons. Thus, it appears that the change in the drainage system occurred within the time interval between 0.79 and 1.60 Ma during the early Pleistocene.

The Teays River abandonment can be further detailed by inspection of the Quaternary glaciation correlation chart of Bowen and others (1986). The first Laurentide glaciation K, at 2.14 Ma, occurred mostly during the two Reunion normal polarity subchrons within the early Matuyama reversed chron. Since the Minford Silt is reversely magnetized, it seems unlikely it was deposited in response to blockage by the K glaciation. It appears more probable that either the pre-Illinoian F or G glaciation during the late Matuyama reversed-polarity chron was responsible for the damming of the Teays River and creation of Lake Tight 0.79 to 0.88 m.y. ago.

Using less detailed paleomagnetic data, Hoyer (1976) came to a broader conclusion: that the drainage change occurred during the time interval between 0.68 and 2.40 Ma, associating the change with either the "Nebraskan" or the "Kansan" glaciations. Paleomagnetic results from southeast Indiana were also found to be reversely polarized (Bleuer, this volume). However, magnetic measurements from ponded sediments of the Old Licking River valley of western Kentucky were determined to be normally magnetized (Teller and Last, 1981). Fullerton (1986) placed the Minford Silt and other members of the Teays Formation in the early Pleistocene with an age greater than 0.788 Ma.

SUMMARY

Paleomagnetic and stratigraphic data indicate that the Minford Silt was deposited during an early Pleistocene glaciation within the Matuyama reversed-polarity chron. The normal polarities in the upper part of the section at Teays Depot were formed by acquisition of a chemical remanent magnetization (CRM) during postdepositional weathering in a normal geomagnetic field, and are not due to a depositional remanent magnetization (DRM) acquired during one of four brief normal polarity subchrons of the Matuyama reversed-polarity chron.

ACKNOWLEDGMENTS

At Marshall University we acknowledge Janet Dudley for her secretarial assistance, Remito Crites for his field and laboratory aid, the Marshall University Sigma Xi Club for a grant for field supplies and equipment, and the Southern Education Research Board for its travel grants (to R.B.B. and D.D.S.). At The Ohio State University, Kevin G. Richardson assisted with the SIRM measurements. An American Chemical Society grant (PRF 9129) (to H.C.N.) and a Marshall University Grant (to R.B.B. and D.D.S.) provided cryogens for liquid helium magnetometer measurements. We also wish to thank Sidney E. White, who rendered valuable suggestions in his review of the manuscript.

REFERENCES CITED

Bloemendal, J., Olfield, R., and Thompson, R., 1979, Magnetic measurements used to assess sediment influx at Lynn Goddiondon: Nature, v. 280, p. 50–53.

Bonnett, R. B., 1975, Analysis of Pleistocene and Holocene drainage changes northwest of the Nitro, West Virginia, region: West Virginia Academy of Science Proceedings, v. 47, no. 3-4, p. 205–211.

Bonnett, R. B., Sanderson, D. D., and Noltimier, H. C., 1978, Paleomagnetic results from the Minford Silt, Teays Depot, West Virginia: EOS Transactions of the American Geophysical Union, v. 59, p. 225.

—— , 1983, Teays Valley stratigraphy and magnetic dating of sediments in West Virginia: Geological Society of America Abstracts with Programs, v. 15, p. 529.

Bowen, D. Q., and 5 others, 1986, Correlation of Quaternary glaciations in the Northern Hemisphere: Quaternary Science Reviews, v. 5, p. 509–510.

Briden, J. C., 1972, A stabiity index of remanent magnetism: Journal of Geophysical Research, v. 77, p. 1401–1404.

Campbell, M. R., 1900, Description of the Huntington Quadrangle, West Virginia: U.S. Geological Survey Geological Atlas, Folio 69.

Cooke, H.B.S., 1973, Pleistocene chronology; Long or short?: Quaternary Research, v. 3, p. 206–220.

Cox, A., 1969, Geomagnetic reversals: Science, v. 163, p. 237–245.

Cox, A., Doell, R. R., and Dalrymple, G. B., 1963, Geomagnetic polarity epochs and Pleistocene geochronometry: Nature, v. 198, p. 1049–1051.

Durrell, R. H., 1961, Pleistocene geology of the Cincinnati area, in Guidebook for field trips; Geological Society of America Annual Meeting, Cincinnati, Ohio: Geological Society America, p. 47–57.

Ettensohn, F. R., 1974, The Pre-Illinoian lake clays of Cincinnati region: Ohio Journal of Science, v. 74, p. 214–226.

Ettensohn, F. R., and Glass, H. D., 1978, Clay-mineral stratigraphy of the pre-Illinoian lake clays from the Cincinnati region: Journal of Geology, v. 86, p. 393–402.

Fisher, R. A., 1953, Dispersion of a sphere: Proceedings of the Royal Society of London, series A., p. 295–305.

Flint, R. F., 1971, Glacial and Quaternary geology: New York, John Wiley and Sons, 892 p.

Fullerton, D. S., 1986, Stratigraphy and correlation of glacial deposits from Indiana to New York and New Jersey: Quaternary Science Reviews, v. 5, p. 23–38.

Goldthwait, R. P., White, G. W., and Forsyth, J. L., 1961, Glacial map of Ohio: U.S Geological Survey Miscellaneous Geological Investigation Map I-316, scale 1:500,000.

Hoyer, M. C., 1976, Quaternary valley fill of the abandoned Teays drainage system of southern Ohio [Ph.D. thesis]: Columbus, Ohio State University, 146 p.

Irving, E., 1967, Evidence for paleomagnetic inclination error in sediment: Nature, v. 213, p. 483–484.

Janssen, R. E., and McCoy, G. P., 1953, Varved clays in the Teays Valley: West Virginia Academy of Science Proceedings, v. 25, p. 53–54.

Leighton, M. M., and Ray, L. L., 1965, Glacial deposits of Nebraskan and Kansas age in northern Kentucky: U.S. Geological Survey Professional Paper 525B, p. B126–B131.

Leverett, F., 1902, Glacial formations and drainage features of the Erie and Ohio basins: U.S. Geological Survey Monograph 41, 802 p.

Lowrie, W., and Alvarez, W., 1981, One hundred million years of geomagnetic polarity history: Geology, v. 9, p. 392–397.

Ray, L. L., 1963, Quaternary events along the unglaciated lower Ohio River Valley, in Geological Survey Research 1963: U.S. Geological Survey Professional Paper 475-B, p. B126–B128.

—— , 1969, Glacial erratics and the problems of glaciation in northeast Kentucky and southeast Ohio; Review and suggestions: U.S. Geological Survey Professional Paper 650-D, p. 195–199.

Richmond, G. M., and Fullerton, D. S., 1986, Introduction to Quaternary glaciations in the United States of America: Quaternary Science Reviews, v. 5, p. 3–10.

Rhodehamel, E. C., and Carlston, C. W., 1963, Geologic history of the Teays Valley in West Virginia: Geological Society of America Bulletin, v. 74, p. 251–274.

Stout, W. E., and Schaaf, D., 1931, Minford Silts of Ohio: Geological Society of America Bulletin, v. 42, p. 663–672.

Swadley, W. C., 1971, The preglacial Kentucky River of northern Kentucky: U.S. Geological Survey Professional Paper 750-D, p. 127–131.

Teller, J. T., 1973, Preglacial (Teays) and early glacial drainage in the Cincinnati area, Ohio, Kentucky, and Indiana: Geological Society of America Bulletin, v. 84, p. 3677–3688.

Teller, J. T., and Last, W., 1981, The Claryville Clay and early glacial drainage in the Cincinnati, Ohio, region: Palaeogeography, Palaeoclimatology, Palaeoecology, v. 33, p. 347–367.

Thornbury, W. D., 1965, Regional geomorphology of the United States: New York, John Wiley and Sons, 609 p.

Thorp, J., and others, 1952, Pleistocene eolian deposits of the United States, Alaska, and parts of Canada: Geological Society of America Map, scale 1:2,500,000.

Tight, W. G., 1903, Drainage modifications in southeastern Ohio and adjacent parts of West Virginia and Kentucky: U.S. Geological Survey Professional Paper 13, 111 p.

Walling, D. E., Peart, M. R., Oldfield, F., and Thompson, R., 1979, Suspended sediment sources identified by magnetic measurements: Nature, v. 281, p. 110–113.

Zijderveld, J.D.A., 1967, AC demagnetization of rocks; Analysis of results, in Collinson, D W., Creer, K. M., and Runcorn, S. K., eds., Methods in Paleomagnetism: New York, Elsevier, p. 254–286.

MANUSCRIPT ACCEPTED BY THE SOCIETY JUNE 29, 1990

Lithology and general stratigraphy of Quaternary sediments in a section of the Teays River Valley of southern Ohio

J. M. Bigham and N. E. Smeck
Department of Agronomy, 2021 Coffey Road, The Ohio State University, Columbus, Ohio 43210

L. D. Norton
Department of Earth and Atmospheric Sciences, Purdue University, West Lafayette, Indiana 47907

G. F. Hall
Department of Agronomy, Ohio State University, Columbus, Ohio 43210

M. L. Thompson
Department of Agronomy, Iowa State University, Ames, Iowa 50011

ABSTRACT

Three transects were conducted across the main channel of the abandoned Teays River valley in Pike, Jackson, and Scioto Counties, Ohio, to evaluate the lithology and general stratigraphy of valley-fill deposits. Field observations obtained from both deep borings and surface excavations indicate that much of the Pleistocene lacustrine fill has been removed and that the modern landscape reflects primarily a sequence of erosional and secondary fill surfaces. Thus, the current valley fill includes lacustrine clays and channel sands, as well as younger sediments of varied origin.

A general sequence of three sedimentary stratigraphic units was commonly encountered in the transects. A silty surface unit overlay an intermediate deposit, which in turn, rested on channel sands or on highly laminated lacustrine materials that were equated with the Minford Clay Member of the Teays Formation. The silty surface unit occurred at almost all sites to a depth of 50 to 60 cm. The mineralogy was mixed, and clay-free particle-size profiles indicated the material was loessial in origin. The intermediate deposit was also encountered in most borings and could be classified as colluvial, alluvial or lacustrine depending on the location. The lithology of this deposit was highly variable and frequently reflected the properties of local bedrock units. Minford Clay was distinguished from younger lacustrine sediments of the intermediate unit on the basis of higher clay content, more micaceous mineralogy, and an elemental Zr content that was two to four times less. Discriminant statistical analyses of data from a total of 180 samples indicated that the Minford Clay could also be easily distinguished from all other Quaternary sediments in the Teays Valley on the basis of selected chemical attributes. By using the same parameters, however, lacustrine deposits in overlying stratigraphic units could not be clearly separated from associated colluvium, alluvium, and loess.

Bigham, J. M., Smeck, N. E., Norton, L. D., Hall, G. F., and Thompson, M. L., 1991, Lithology and general stratigraphy of Quaternary sediments in a section of the Teays River Valley of southern Ohio, *in* Melhorn, W. N., and Kempton, J. P., eds., Geology and hydrogeology of the Teays-Mahomet Bedrock Valley System: Boulder, Colorado, Geological Society of America Special Paper 258.

INTRODUCTION

The abandoned Teays drainage system in southern Ohio, West Virginia, and Kentucky has been mapped and studied in detail by numerous individuals for various purposes over the past 150 years. The rich geologic history of the Teays is reflected in extensive, unconsolidated, valley-fill deposits that choke both the main valley and its many tributaries. These deposits include original Teays-age alluvial sediments and Pleistocene lacustrine material in association with younger erosional and depositional debris (Hoyer, 1976). As a result, modern landform, sediment, and soil relations in the Teays Valley are complex.

The relatively level topography of the main Teays channel in southern Ohio has traditionally made it an attractive agricultural area, and in recent years, personnel from the U.S. Department of Agriculture Soil Conservation Service, the Ohio Department of Natural Resources, and The Ohio State University have been involved in a cooperative inventory of soil resources in the valley. As part of this effort, detailed studies of the properties of both soils and surficial sediments in the valley have been conducted. Results pertaining to the evolution of soils in the Teays Valley have been described elsewhere (Thompson and others, 1981; Thompson and Smeck, 1983). Our objective here is to summarize data relevant to the chemical, physical, and mineralogic characteristics of surficial deposits in the main channel of the Teays. It is hoped that this information will supplement existing knowledge pertaining to the characteristics of Teays valley fill and provide additional insight into the complex geologic history of the area.

STUDY AREA

The area examined in this study consisted of a 50-km unglaciated segment of the main Teays channel between Minford and Waverly, Ohio (Fig. 1). Throughout this section, the valley is 1.25 to 2.5 km wide, is bounded by fairly steep bedrock walls, and has a maximum relief (from the uplands to the present valley floor) of about 90 m. The valley floor is generally sloping or undulating, suggesting a predominantly erosional topography. In places, resistant bedrock shelves or rock terraces jut into the valley, but whether or not they constitute true cut terrace levels of the Teays River is not known.

Field studies were concentrated along three transects of the main valley in Pike (PK) (Sec. 34,T.5N,R.21W; Secs. 1 and 12,T.4N,R.21W), Jackson (JK) (Sec. 30,T.6N,R.19W; Secs. 25 and 36,T.5N,R.20W), and Scioto (SC) Counties (Secs. 17 and 18,T.4N,R.20W) (Fig. 1); transect locations were selected to include representative units of the major geomorphic surfaces currently expressed in this segment of the valley. In addition, a 9-m profile of undisturbed Minford Clay was described and sampled from an erosional remnant located near the middle of the valley floor along Taylor Hill Road (Sec. 3,T.3N,R.20W), approximately 2.7 km southeast of the type locality in the community of Minford. The type section was avoided due to its long exposure and partial burial by colluvial debris.

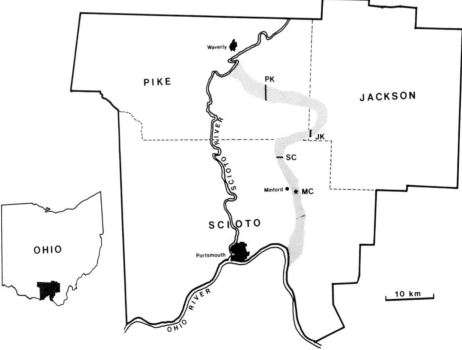

Figure 1. Study area in Pike, Jackson, and Scioto Counties, Ohio, with transects denoted as PK, JK, and SC, respectively. MC refers to the Minford Clay section.

FIELD AND LABORATORY PROCEDURES

In the initial stages of this investigation, detailed soil maps were completed in each of the three transect areas. Subsequently, a series of 2.5-cm-diameter cores was obtained to depths of up to 9 m using a truck-mounted hydraulic probe. Coring sites were established on major geomorphic surfaces and were spaced at intervals of 150 to 600 m across the valley floor. Both the nature and thickness of valley-fill deposits were noted at each location. Based on these observations, four sites were selected along each transect to represent typical landscape positions and the range of surficial sediments present in the valley (Fig. 2). At each site, soil profiles were described, and bulk (~5 kg) samples were obtained from backhoe trenches excavated to depths of 2 to 3 m. All samples corresponded to soil horizons and/or major stratigraphic units. At the Minford Clay site (marked MC in Fig. 1), samples were obtained from a continuous 9-m core obtained with the hydraulic probe using a 2.5-cm open-face tube. The core was subsequently described and sampled in approximately 15-cm increments.

Samples were air-dried, ground between wooden rollers, sieved to remove material >2 mm, and thoroughly mixed. Particle-size analyses of the <2-mm materials were conducted using a modified sedimentation-pipette procedure (Steele and Bradfield, 1934) with sodium hexametaphosphate as a dispersing agent. Calcium carbonate equivalents (calcite + dolomite) were determined with a Chittick apparatus (Dreimanis, 1962).

Forty-gram subsamples were separated into sand (2,000 to 50 μm), silt (50 to 2 μm), and clay (<2 μm) fractions using standard sieve and gravity sedimentation techniques (Jackson, 1975; Rutledge and others, 1967). Prior to fractionation, carbonates (if present) were dissolved with a 1N acetic acid–sodium acetate buffer (pH 5) and organic compounds were removed with H_2O_2. Silt fractions were subsequently analyzed by x-ray fluorescence for total Ti, Zr, and K contents using samples prepared as indicated by Rutledge and others (1975b). Total K in the <2-μm materials was determined by atomic absorption analysis following complete dissolution of the samples in a Teflon-lined acid digestion bomb using the method of Bernas (1968).

Clay mineralogy data were obtained by x-ray diffraction analysis of 30-mg samples of clay oriented on 25 × 45-mm glass slides. All specimens were x-rayed using Cu-Kα radiation with a diffraction assembly that included a theta-compensating slit, a medium-resolution receiving slit, and a diffracted beam monochromator. Standard chemical (Mg-saturation, K-saturation, ethylene glycol solvation) and thermal (25°, 350°, 550°C) pretreatments were employed. Quantitative determinations of kaolinite were also obtained from selected clays using differential scanning calorimetry with standards prepared as indicated by Dixon (1966).

Finally, discriminant statistical analyses of the laboratory data were performed to test the utility of various parameters for differentiating sedimentary units observed in the field (Howarth, 1971; Hawkins and Rasmussen, 1973; Paton and Little, 1974;

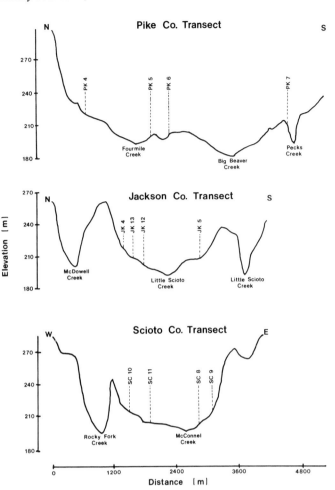

Figure 2. Topographic sections for the Pike, Jackson, and Scioto County transects. 20× vertical exaggeration.

Norton and Hall, 1985). All calculations and tests were conducted using established programs of the Statistical Analysis System (SAS Institute, 1979) and the Biomedical Data Programs (BMDP, 1979).

RESULTS AND DISCUSSION

Minford Clay

Stout and others (1943), in a mid-century description of the Teays drainage system, noted that the most conspicuous deposits in the Teays Valley and its tributaries were the plastic, highly laminated Minford "silts." Although the Minford material is actually a clay or silty clay rather than a silt (Hoyer, 1976), it is, nonetheless, one of the most striking remnants of the Teays period depositional environment. The Minford Clay (terminology of Hoyer, 1976) and contemporaneous lacustrine materials have been widely recognized and correlated throughout the Teays drainage system (e.g., Rhodehamel and Carlston, 1963; Etten-

TABLE 1. PROPERTIES AND VARIABILITY OF THE MINFORD CLAY AT SITE MC

Property (%-wt/wt)	Brown Clay (0.1–6.9 m)				Gray Clay (6.9–9.2 m)			
	n	X	Range	CV (%)	n	X	Range	CV (%)
Clay	41	71.9	59.1–86.3	9.4	14	79.6	74.3–86.7	5.6
Silt	41	27.9	13.7–40.9	24.0	14	20.4	13.3–25.7	22.3
CCE*	41	4.5	1.3–9.8	50.2	14	3.6	1.3–6.1	43.3
K_2O†	17	5.3	4.0–5.5	6.4	7	5.1	4.5–5.6	7.6
K§	17	3.6	2.8–4.3	9.1	7	3.5	3.0–4.0	9.1
Ti§	17	0.82	0.77–0.86	3.6	7	0.79	0.76–0.81	2.6
Zr§	17	0.014	0.012–0.021	18.9	7	0.017	0.014–0.022	23.7

*CCE = calcium carbonate equivalent.
†<2 µm fraction.
§2–50 µm fraction.

sohn and Glass, 1978); however, these materials have been subjected to relatively few detailed physical, chemical, and mineralogical analyses. Hence, our purpose in describing and analyzing a major section of the Minford Clay was to supplement existing knowledge and to provide a standard for comparison with similar materials encountered elsewhere in the valley.

The Minford materials examined in this study were obtained from an erosional remnant that forms a small hill near the middle of the valley floor. The top of the hill occurs at an elevation of 216 m and appears to coincide with a major erosional surface (213 to 220 m) expressed throughout the main channel, especially along the valley walls. Except for a thin, 13-cm loess cap, the entire section consisted of rhythmitic clays and silty clays (Table 1) with laminae ranging from 1 to 2 mm in thickness; sand contents were <0.2 percent throughout. Although we were unable to penetrate beyond a depth of 9.5 m with our coring device, topographic features suggest the clays extend to a depth of at least 15 m. At a nearby location, Hoyer (1976) observed 14.5 m of clay.

The entire column showed no to mild effervescence with 10 percent HCl. Laboratory calcium carbonate equivalents (calcite + dolomite) range from 1.3 to 9.8 percent (Table 1). Although a zone of pedogenic carbonate accumulation occurred near the base of the modern soil solum (at an approximate depth of 72 cm), there appeared to be no systematic stratification of the clay deposit with respect to carbonates. Abundant gypsum crystals of up to 0.5 cm in diameter were observed between the depths of 1.25 and 2.0 m; confirmation of their mineralogy was obtained by x-ray analysis of powdered specimens.

In the upper 6.9 m of the section, all clays were brown to yellowish brown (10YR 5/3 to 10YR 5/4) with occasional bluish gray (5B 5/1) mottles and brownish yellow (10YR 6/8) coatings of goethite. At 6.9-m depth there was an abrupt transition to a uniform, dark gray (10YR 4/1 or 5Y 4/1) clay. Similar occurrences of "oxidized" brown clay overlying "unoxidized" gray clay have previously been described in Minford deposits (Stout and others, 1943; Hoyer, 1976). In Table 1, we have adopted Hoyer's (1976) nomenclature and have termed these segments of the section "brown" and "gray" clay. At this location, the gray material appears to be somewhat enriched in clay relative to the overlying brown zone; however, both average more than 70 percent clay. The micaceous nature of the sediments is reflected in average total K contents of 4.3 and 3.6 percent for the clay and silt fractions, respectively (Table 1). In similar analyses of unfractionated samples of Minford "silt" collected from three locations in the valley, Stout and others (1943) obtained total K contents ranging from 2.7 to 4.0 percent.

The clay mineralogy of the brown and gray Minford samples examined in this study is summarized in Table 2 and is based on the combined results of chemical, x-ray, and thermal analyses. If 1 percent K_2O can be equated with 10 percent clay mica, as reported by Jackson (1975), then the mica (illite) content of the clay fraction ranges from 40 to 55 percent throughout the section, with no significant difference between the brown and gray clays. These mica values are less than those obtained by Hoyer (1976) from Minford samples collected in the main Teays channel, are more than those reported by Rhodehamel and Carlston (1963) for two West Virginia samples, and are comparable to the results of Ettensohn and Glass (1978) for Teays-age clays from two locations in the Cincinnati area. In our case, x-ray diffraction analyses confirm a uniform clay mica content for the brown and gray clays and further indicate that the remainder of the clay mineral suite is in both cases dominated by kaolinite and vermiculite with small amounts of associated quartz and interstratified clay mica-vermiculite. No smectite was observed. Whereas the clay mineralogy of both materials is quite similar at this location, thermal analyses indicate that the kaolinite content of the gray clay is consistently greater (by ~10 percent) than in the overlying brown material.

Elemental Ti and Zr contents and, especially, ratios of Ti to Zr in "immobile" size fractions have been widely used to locate or confirm lithologic discontinuities in soils and unconsolidated

TABLE 2. CLAY MINERALOGY OF BROWN AND GRAY UNITS OF THE MINFORD CLAY AT SITE MC

No. of Samples	Material	Mineral Component				
		Clay Mica (%)	Kaolinite (%)	Vermiculite (%)	Inter stratified* (%)	Quartz (%)
17	Brown clay	53	20	22	5	trace
7	Gray clay	51	30	14	5	trace

*Randomly interstratified clay mica-vermiculite.

TABLE 3. TYPE AND THICKNESS OF SURFICIAL DEPOSITS OBSERVED AT DETAILED STUDY SITES IN PIKE, JACKSON, AND SCIOTO COUNTIES

Site	Material				
	Loess	Colluvium	Alluvium	Lacustrine	Residuum
	Depth (cm)				
PK-4	0–48	48–61	...	61–241+	...
PK-5	0–51	...	...	51–211+	...
PK-6	0–56	...	56–145	145–170+	...
PK-7	0–51	...	51–185+	...	...
JK-4	0–66	66–170	...	170–254+	...
JK-5	0–41	41–102	...	...	102–173+
JK-12	0–76	76–201	201–254+	...	...
JK-13	0–56	56–180	180–274+	...	...
SC-8	0–51	51–201	...	201–226+	...
SC-9	0–43	43–127	...	127–231+	...
SC-10	0–46	46–147	...	...	147–160+
SC-11	0–25	...	...	25–183+	...

sediments (e.g., Wilding and others, 1971). Values of Ti and Zr obtained from the total silt fractions of the Minford core (Table 1) are relatively uniform throughout, and their ratio varies unsystematically between 40 and 65 (not shown), with no indication of an unconformity between the brown and gray clays. This result could be interpreted to mean that the brown clays have undergone oxidation in situ; however, the lack of a distinct mineral weathering front and the impermeable nature of the clays suggest that the brown clays were initially deposited in an oxidized state, as Ettensohn (1974) concluded for comparable Cincinnati-area deposits.

General stratigraphy of Teays valley sediments

Hoyer (1976) proposed a general, vertical stratigraphic sequence for the Teays valley fill consisting of basal sand or silt, overlain by Minford Clay, with a capping of highly variable loamy sediments that include loess, colluvium, and recent alluvium. In Ohio, the basal sand unit is frequently referred to as the "Gallia Sand" and is thought to consist of both preglacial stream alluvium and deltaic deposits constructed as proglacial Lake Tight (Tight, 1903) was formed (Norris and Spicer, 1958). Since both the Gallia Sand and the Minford Clay were deposited prior to or during the Lake Tight period in the Teays Valley, they are commonly grouped and classified as the Teays Formation.

Field observations suggest that erosional downcutting since the time of Lake Tight has extended below the level of the original bottom of Lake Tight throughout the valley. The present topography is predominantly erosional with no well-defined depositional surfaces. Topographic sections from the transect areas (Fig. 2) suggest the presence of at least three surfaces cut into the current valley fill. The youngest and lowest surface consists of the slopes to modern drainageways below 200 m. A second appears to occur at the 200 to 210 m level and corresponds to detailed sampling sites PK-5 and 6, JK-5, 12, and 13, and SC-8 and 11. The third and highest surface appears to occur at an elevation of 213 to 220 m (PK-4 and 7, JK-4, SC-9 and 10). No differences in soil development were apparent among the suspected geomorphic surfaces, and more field work is needed to firmly establish their existence. At present, there also appears to be no well-defined relation between the proposed surfaces and the general stratigraphy of Teays valley fill.

In transects of the main Teays channel, at least five types of unconsolidated sediments were frequently encountered. These included loess, colluvium, alluvium (both recent and old), lacustrine sediments, and residuum from the local sedimentary rocks. Table 3 provides a summary of the stratigraphic units identified at each of the 12 detailed study sites in Pike, Jackson, and Scioto Counties. Although the vertical sections analyzed at these sites were all less than 3 m deep, the materials examined were representative of major stratigraphic units observed in deeper borings throughout the valley. Thompson and others (1981) have provided a schematic cross section of the valley at the Jackson County transect that illustrates the general relation of the major stratigraphic units.

A silty surface unit, ranging from 25 to 75 cm in thickness, was encountered at all the sites examined in this study. Because of its silty texture, bright color (usually 10YR 5/6), and consistent depth across the landscape, this material was identified as loess. Thompson and others (1981) calculated clay-free particle-size profiles for the Jackson County sections and found that medium silt was the dominant particle-size fraction in the surface mantle. The loess depth at the sites in this study is essentially the same as that reported by Rutledge and others (1975a) on the leeward side of a transect of Scioto River loess. Rutledge and others (1975a) concluded that the surficial loess in their study was Late Wisconsinan. Because our transects are approximately the same distance east of the Scioto River and only 15 to 25 km south of the transect of Rutledge and others (1975a), we believe that the loess mantle in the Teays Valley is contemporaneous in age. The limited degree of soil development in the loess mantle also supports a

Late Wisconsinan age, and Thompson and others (1981) have presented evidence to indicate that this loess mantled a land surface that often included a well-developed paleosol. Prior to shallow burial, this paleosol had formed in a variety of sedimentary materials with variable lithology.

At 10 of the 12 detailed study sites (Table 3), colluvium and/or alluvium were identified as materials subjacent to the loess. In some instances, especially along the valley walls (e.g., JK-4 and SC-9), the colluvium displayed classic features, such as stone lines, drag folds, heterogeneous particle-size distributions, and fragments of local rocks from upslope positions. Near the valley floor, materials were frequently encountered with characteristics of both colluvium and alluvium and which were probably deposited through the combined action of mass wasting and slope wash. Thompson and Smeck (1983) have cited micromorphologic evidence in the form of banded plasma and silt particles to suggest that loamy and silty colluvium at sites JK-12 and 13 may have moved downslope during periods of periglacial solifluction. Older, coarse-textured alluvium was also identified below the more recent colluvium at these sites and was correlated with the Gallia Sands.

Lacustrine units were described at 7 of the 12 detailed study areas (Table 3), in site positions ranging from the steeply sloping channel walls to the nearly level valley floor. No consistent relations existed with regard to current valley landforms, suggesting that a tremendous volume of sediment was removed during and after the drainage of Lake Tight. All the lacustrine deposits observed in the field were laminated with textures dominated by silt and clay, and initially, all such laminated sediments were correlated with the Minford Clay. However, laboratory analyses indicate that at least two major lacustrine sections are preserved in this segment of the Teays Valley, only one of which possesses lithologic characteristics similar to those of Minford Clay.

Lithology of the Teays valley fill

Physical and chemical properties of sediments from the detailed study sites in Pike, Jackson, and Scioto Counties are summarized in Table 4. Due to the variability and overlapping characteristics of colluvial and alluvial units in the valley, these materials have been grouped for discussion purposes.

The uniform nature of the thin loess mantle encountered at all deep boring and detailed study sites in the valley is reflected in low coefficients of variation for the physical and chemical parameters evaluated (Table 4). Clay contents in the loess show the greatest variability. Textures range from silt loams to silty clay loams, but are within the range commonly encountered in loessial materials subjected to soil formation. The clay mineralogy of the loess is decidedly mixed, with average clay mica, Al-interlayered vermiculite-smectite, kaolinite, and quartz contents of 20, 60, 15, and 5 percent, respectively (data not shown). Ti/Zr ratios for the total silt fractions range between values of 10 to 15 and are uniform both within and among sites (e.g., Fig. 3).

In contrast, the colluvial/alluvial units, due in part to the nature of the materials and in part to our grouping, are highly variable. Mineralogical analyses of these materials frequently reflect the lithology of local Mississippian sandstones and shales that crop out along the valley walls. Occasional encounters with bedrock and/or residuum in transects of the valley bottom (e.g., Site JK-5, Table 3) also suggest that in some reaches the modern streams have cut through the fill to points close to the level of the bedrock floor.

As noted previously, laminated lacustrine sediments were present at 7 of the detailed study sites. In some cases, the physical, chemical, and mineralogic properties of these sediments are virtually identical to those obtained from the Minford Clay materials at site MC (Tables 1, 4); that is, they are clayey, micaceous, and extremely low in elemental Zr. Other lacustrine units can be

TABLE 4 PROPERTIES AND VARIABILITY OF SURFICIAL DEPOSITS (0–2.5 M) OBSERVED AT DETAILED STUDY SITES IN PIKE, JACKSON, AND SCIOTO COUNTIES

Property (% wt/wt)	Material	n	X	Range	CV (%)
Clay	Loess	40	22.1	12.4–35.8	27.7
	Colluvium/Alluvium	68	29.0	10.8–66.1	34.4
	Lacustrine (post-Minford)	27	36.4	18.9–64.2	35.5
	Lacustrine (Minford)	15	73.5	64.3–84.3	8.9
	Residuum	6	25.8	12.1–31.5	28.1
Silt	Loess	40	65.9	50.4–79.0	9.7
	Colluvium/Alluvium	68	46.9	12.2–65.4	28.9
	Lacustrine (post-Minford)	27	56.0	29.4–70.8	20.2
	Lacustrine (Minford)	15	25.4	15.5–35.2	27.7
	Residuum	6	38.7	20.6–69.7	53.3
K_2O*	Loess	24	2.1	1.6–2.9	14.1
	Colluvium/Alluvium	34	2.4	1.7–4.0	22.9
	Lacustrine (post-Minford)	27	2.7	2.0–3.5	14.3
	Lacustrine (Minford)	10	5.0	4.4–5.2	5.1
	Residuum	...	...	...	...
K†	Loess	30	1.5	1.0–1.8	12.9
	Colluvium/Alluvium	51	1.3	0.47–2.5	29.9
	Lacustrine (post-Minford)	27	1.9	1.5–2.4	12.5
	Lacustrine (Minford)	12	3.1	2.7–3.3	5.4
	Residuum	5	1.7	1.2–2.4	32.4
Ti†	Loess	30	0.73	0.65–0.80	5.9
	Colluvium/Alluvium	51	0.83	0.65–0.93	7.6
	Lacustrine (post-Minford)	27	0.84	0.73–0.96	7.5
	Lacustrine (Minford)	12	0.84	0.79–0.97	2.7
	Residuum	5	0.88	0.82–0.92	4.6
Zr†	Loess	30	0.062	0.047–0.076	11.5
	Colluvium/Alluvium	51	0.070	0.036–0.219	36.9
	Lacustrine (post-Minford)	27	0.047	0.023–0.084	30.8
	Lacustrine (Minford)	12	0.019	0.014–0.026	19.2
	Residuum	5	0.099	0.043–0.142	50.4

*<2 μm fraction.
†2–50 μm fraction.

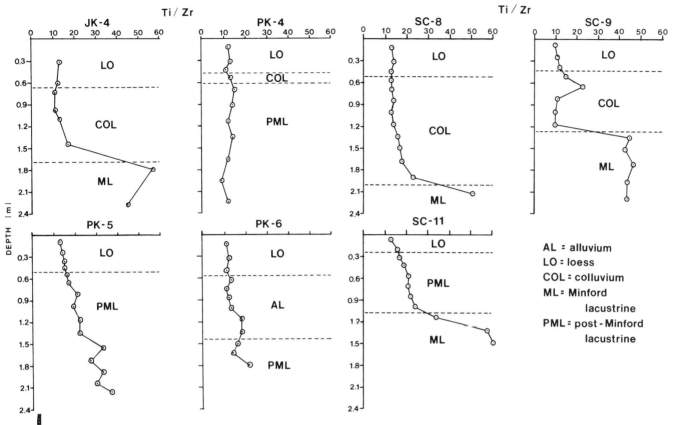

Figure 3. Ti/Zr profiles (2- to 50-μm fractions) for detailed study sections containing lacustrine sediments.

easily separated on the basis of laboratory parameters (Table 4). Multiple lacustrine units were only encountered in one study section (SC-11). Since the Minford Clay underlies the "non-Minford" sediment at this location, the latter is hereafter referred to as "post-Minford." Verification of the relative ages of these units will require further study but should be facilitated by the fact that their lithologic characteristics are distinctly different. With the exception of elemental Ti, the ranges of all physical and chemical properties presented in Table 4 are, in fact, mutually exclusive for these materials. Generally, the Minford clay is uniform in character, whereas the post-Minford lacustrine deposits are more variable.

Ratios of elemental Ti and Zr in the silt fraction provide an excellent means of distinguishing the Minford Clay from the post-Minford lacustrine sediments and other overlying materials (Fig. 3). Ti/Zr ratios range between 40 and 65 and, due to the low Zr content of the Minford Clay, are consistently greater (by as much as six times) than ratios obtained from associated loess, alluvium, colluvium, or lacustrine units. The strikingly lower Zr content of the Minford Clay as compared to the other materials suggests different source regions. It is probable that the Minford sediments originated south of the study area toward the headwaters of the Teays River in the Piedmont Plateau of Virginia and North Carolina; whereas, the overlying materials originated locally or from northern glacial sources. In some cases (e.g., PK-5), ratios from the post-Minford lacustrine sediments show a gradual increase with depth, suggesting that the original Minford Clay may have been reworked in local cycles of erosion and sedimentation. The sensitivity of the elemental analyses to the presence of Minford Clay is represented by the Ti/Zr profile for section SC-9. At this site, a tongue of the basal Minford Clay extended upward into the overlying colluvium and was inclined in a downslope direction like a miniature drag fold. This feature is reflected in a bulge of the Ti/Zr profile at a depth of approximately 65 cm (Fig. 3). At the scale utilized in Figure 3, the Ti/Zr ratios are not very useful for distinguishing lithologic discontinuities among materials overlying the Minford Clay.

The chemical and physical data from a total of 180 samples collected in the Teays Valley were subjected to a discriminant function analysis to determine if the various stratigraphic units could be separated on a statistical basis. The procedure employed was similar to that utilized by Paton and Little (1974) for analyzing a valley-fill sequence in eastern Australia. By using this approach, the various lithologic parameters in Table 4 were initially analyzed for their ability to distinguish stratigraphic units. Based on results from a multivariate analysis, total K_2O content of the clay fractions, and total Ti and total Zr in the silt fractions proved to be the three most sensitive variables. Approximately 92

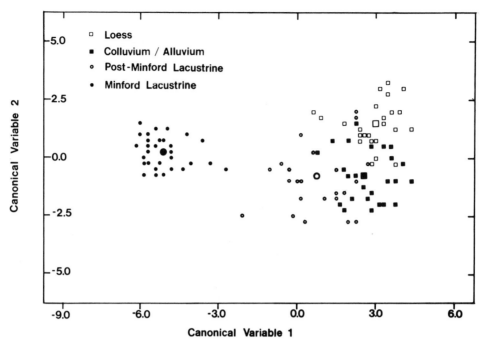

Figure 4. Scatter plot produced by a stepwise discriminant analysis of lithologic data from Teays valley-fill deposits using scores for the first two canonical variables. Large symbols represent means for each group.

percent of the total dispersion in the data was accounted for by K_2O alone, 98 percent by K_2O and Ti together, and 100 percent by the combination of K_2O, Ti, and Zr.

By using these three parameters, a stepwise discriminant function was subsequently developed (Rao, 1970; BMDP, 1979) to test the field and laboratory classifications of sedimentary units in the Teays Valley. Figure 4 is a scatter plot for the cases examined using scores from the following two canonical variables:

canonical variable 1 = 6.79 − 2.06 K_2O − 2.49 Ti + 40.31 Zr
canonical variable 2 = 19.78 − 0.21 K_2O − 21.98 Ti − 26.58 Zr.

As suggested by this plot, the statistical function correctly classified 100 percent of the Minford Clay samples, but could not clearly discriminate the other stratigraphic units. This result suggests that the post-Minford lacustrine units are more closely related to "modern" valley-fill deposits than to the early Teays-age Minford Clay.

CONCLUSIONS

A general sequence of three sedimentary stratigraphic units was commonly encountered in transects of the main Teays channel in Pike, Jackson, and Scioto Counties, Ohio. A ubiquitous silt mantle overlay an intermediate unit, which in turn, rested on highly laminated lacustrine materials that were equated with the Minford Clay Member of the Teays Formation.

The silty surface unit was present at almost all sites to a depth of 50 to 60 cm and was recognized as Wisconsinan loess, based on its distribution, thickness, and similarities to the documented regional occurrence of Wisconsinan loess by Rutledge and others (1975a). The intermediate deposit was also encountered in most borings and could be classified as colluvial, alluvial, or lacustrine depending on the location. The lithology of this deposit was highly variable and frequently reflected the mineralogy of local bedrock units.

Minford Clay was distinguished from younger lacustrine sediments of the intermediate unit on the basis of greater clay content, more micaceous mineralogy, and an elemental Zr content that was two to four times lower. Discriminant statistical analyses of data from a total of 180 samples indicated that the Minford Clay could also be easily distinguished from all other Quaternary sediments in the Teays Valley on the basis of its chemical attributes. In contrast, lacustrine deposits in the overlying stratigraphic units could not be clearly separated from associated colluvium, alluvium, and loess by using the same parameters.

ACKNOWLEDGMENTS

This study could not have been completed without the assistance and cooperation of numerous field and office personnel affiliated with the U.S. Department of Agriculture Soil Conservation Service (USDA-SCS), the Ohio Department of Natural Resources (ODNR) and The Ohio State University (OSU). We gratefully acknowledge the ideas and efforts of the following individuals: E. Gamble, R. Gehring, R. Hendershot, K. Huffman, R. Mapes, and L. Tornes of the USDA-SCS: R. Christman, M. Fuesner, S. Hamilton, L. Jones, F. McCleary and A. Ritchie of the ODNR; K. Brady, W. Jaynes, E. Long, M. Ransom, and S. Shipitalo of OSU.

Contribution from Department of Agronomy, The Ohio State University, Columbus. Salaries and research support provided by state and federal funds appropriated to the Ohio Agricultural Research and Development Center, The Ohio State University.

REFERENCES CITED

Bernas, B., 1968, A new method for decomposition and comprehensive analysis of silicates by atomic absorption spectrometry: Analytical Chemistry, v. 40, p. 1682–1686.

BMDP, 1979, Dixon, W. J., and Brown, M. B., eds., Biomedical computer programs P-series, Los Angeles, University of California: Berkeley, University of California Press.

Dixon, J. B., 1966, Quantitative analysis of kaolinite and gibbsite in soils by differential thermal and selective dissolution methods: Clays and Clay Minerals, v. 14, p. 83–89.

Dreimanis, A., 1962, Quantitative gasometric determination of calcite and dolomite by using the Chittick apparatus: Journal of Sedimentary Petrology, v. 32, p. 520–529.

Ettensohn, F. R., 1974, The pre-Illinoian lake clays of the Cincinnati region: Ohio Journal of Science, v. 74, p. 214–226.

Ettensohn, F. R., and Glass, H. D., 1978, Clay-mineral stratigraphy of pre-Illinoian lake clays from the Cincinnati region: Journal of Geology, v. 86, p. 393–402.

Hawkins, D. M., and Rasmussen, S. E., 1973, Use of discriminant analysis for classification of strata in sedimentary successions: Mathematical Geology, v. 5, p. 163–177.

Howarth, R. J., 1971, An empirical discriminant model applied to sedimentary rock classification from major element geochemistry: Mathematical Geology, v. 3, p. 51–60.

Hoyer, M. C., 1976, Quaternary valley fill of the abandoned Teays drainage system in southern Ohio [Ph.D. thesis]: Columbus, Ohio State University (Diss. abs. 77-2417), 163 p.

Jackson, M. L., 1975, Soil chemical analysis; Advanced course, 2nd ed.: Madison, Wisconsin, privately published.

Norris, S. E., and Spicer, H. C., 1958, Geological and geophysical study of the preglacial Teays valley in west-central Ohio: U.S. Geological Survey Water Supply Paper 1460-E, p. 199–232.

Norton, L. D., and Hall, G. F., 1985, Differentiation of lithologically similar soil parent materials: Soil Science Society of America Journal, v. 49, p. 409–414.

Paton, T. R., and Little, I. P., 1974, Discriminant-function analyses on a valley-fill sequence in southeastern Queensland: Geoderma, v. 11, p. 29–36.

Rao, C. R., 1970, Advanced statistical methods in biometrics: New York, John Wiley & Sons, p. 236–272.

Rhodehamel, E. C., and Carlston, C. W., 1963, Geologic history of the Teays Valley in West Virginia: Geological Society of America Bulletin, v. 74, p. 251–274.

Rutledge, E. M., Wilding, L. P., and Elfield, M., 1967, Automated particle-size separation by sedimentation: Soil Science Society of America Proceedings, v. 31, p. 287–288.

Rutledge, E. M., Holowaychuk, N., Hall, G. F., and Wilding, L. P., 1975a, Loess in Ohio in relation to several possible source areas; 1. Physical and chemical properties: Soil Science Society of America Proceedings, v. 39, p. 1125–1132.

Rutledge, E. M., Wilding, L. P., Hall, G. F., and Holowaychuk, N., 1975b, Loess in Ohio in relation to several possible source areas; 2. Elemental and mineralogical composition: Soil Science Society of America Proceedings, v. 39, p. 1133–1139.

SAS Institute, 1979, SAS user's guide: Raleigh, North Carolina, SAS Institute, 494 p.

Steele, J. G., and Bradfield, R., 1934, The significance of size distribution in the clay fraction: American Soil Survey Association Bulletin, v. 15, p. 88–93.

Stout, W. E., Ver Steeg, K., and Lamb, G. F., 1943, Geology of water in Ohio: Ohio Geological Survey, 4th series, Bulletin 44, 694 p.

Thompson, M. L., and Smeck, N. E., 1983, Micromorphology of polygenetic soils in the Teays River Valley, Ohio: Soil Science Society of America Journal, v. 47, p. 734–742.

Thompson, M. L., Smeck, N. E., and Bigham, J. M., 1981, Parent materials and paleosols in the Teays River Valley, Ohio: Soil Science Society of America Journal, v. 45, p. 918–925.

Tight, W. G., 1903, Drainage modifications in southeastern Ohio and adjacent parts of West Virginia and Kentucky: U.S. Geological Survey Professional Paper 13, 111 p.

Wilding, L. P., Drees, L. R., Smeck, N. E., and Hall, G. F., 1971, Mineral and elemental composition of Wisconsin-age till deposits in west-central Ohio, in Goldthwait, R. P., ed., Till; A symposium: Columbus, Ohio State University Press, p. 290–317.

MANUSCRIPT ACCEPTED BY THE SOCIETY JUNE 29, 1990

Printed in U.S.A.

The Old Kentucky River; A major tributary to the Teays River

James T. Teller
Department of Geological Sciences, University of Manitoba, Winnipeg, Manitoba R3T 2N2 Canada
Richard P. Goldthwait
P.O. Box 656, Anna Maria, Florida 34216

ABSTRACT

The Old Kentucky River system was a major contributor to the Teays River, draining southwestern Ohio and much of eastern Kentucky. The trunk river flowed northward from southeastern Kentucky throughout Frankfort and Carrollton, and then past Cincinnati and Dayton, joining the Teays River near Springfield, Ohio. North of the glacial boundary, which lies along the modern Ohio River, the course of the Old Kentucky River has been modified, and is today largely buried by drift. Although dissection is extensive to the south, there are many remnants of this entrenched and broadly meandering Teays-age valley system and of its sub-upland predecessors. These valleys contain areas of upward-fining, deeply weathered gravel, composed mainly of rounded quartz, chert, and silicified limestone pebbles derived from the headwaters of the system. Modern rivers have been entrenched 30 to 100 m below the Old Kentucky River valley and its main tributaries, the Old Licking and South Fork.

The Old Kentucky River system was severed from the Teays when glaciation dammed its downstream reaches, forcing a reversal in flow direction between its junction with the Teays in west-central Ohio and Carrollton, Kentucky, and causing westward overflow into the Old Ohio River system. Piracy by the Old Ohio may also have contributed to the integration of the Old Kentucky and Old Ohio River basins. Ponded sediment is present in some of the now-abandoned valley remnants east of Cincinnati. As a result of glacial damming, the headwaters of the Teays River in southeastern Ohio and West Virginia overflowed westward across the Manchester divide into the Old Kentucky River drainage basin. All of these events led to establishment of the modern Ohio River system.

INTRODUCTION

Regional setting

The major preglacial river system in the east-central part of the United States was the Teays, first studied in detail by Tight (1903) and named for now-abandoned valleys in West Virginia. This ancient drainage system headed in the Appalachians, and the trunk stream, the Teays River, flowed westward across central Ohio, Indiana, and Illinois to the Mississippi Embayment (Fig. 1). Much of the old Teays River system is now buried beneath glacial drift, and is recognized only in boreholes. Beyond the margin of glaciation (Fig. 1), in southeastern Ohio, West Virginia, and Kentucky, remnants of the headwaters of this ancient drainage system are exposed at the surface. These old valleys have a meandering pattern, and lie at elevations above present-day rivers that have deeply eroded their ancient floors and have even breached divides between old watersheds.

The Old Kentucky River was the largest tributary to the Teays River, extending from southern Kentucky through Frankfort, north to Cincinnati and Dayton, Ohio, and then northeastward to about Springfield, where it joined the Teays. Major

Teller, J. T., and Goldthwait, R. P., 1991, The Old Kentucky River; A major tributary to the Teays River, *in* Melhorn, W. N., and Kempton, J. P., eds., Geology and hydrogeology of the Teays-Mahomet Bedrock Valley System: Boulder, Colorado, Geological Society of America Special Paper 258.

tributaries, such as the Old Licking River, which headed in eastern Kentucky, joined the Old Kentucky River near Cincinnati (Fig. 1).

The Old Kentucky River and its tributaries cut their valleys mainly in the flat-lying interbedded limestones and shales of Late Ordovician age on the Cincinnati Arch. Only in the headwaters of the ancient river system—south of where distinct valley remnants still exist, and several hundred kilometers upstream from Cincinnati—do younger Silurian and Devonian carbonates and shales and Mississippian and Pennsylvanian clastic rocks form the regional bedrock. The distinctive nature of some of these lithologies, such as those that contain silicified Mississippian corals and bryozoans, rounded Pennsylvanian quartz pebbles, and coal, allows them to be traced downstream in parts of the Old Kentucky River system as far as the glacial boundary.

The demise of the great Teays River system was caused by glaciation, which dammed and then filled its trunk valley a number of times, forcing those rivers beyond the glacial margin to overflow and establish new channels. This may have been a gradual process, occurring step by step, as first one and then another continental ice sheet invaded the Teays River watershed. Piracy by the Old Ohio River near Madison, Indiana, probably played a major role in diverting waters of the Old Kentucky River from the Teays system to the west.

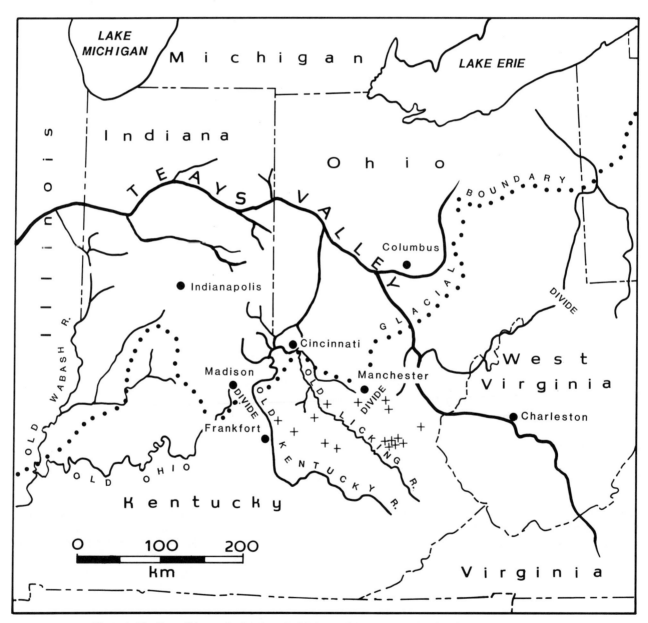

Figure 1. The Teays River and related preglacial rivers of the east-central U.S. (after Thornbury, 1965; Luft, 1980; Burger and others, 1966; Wayne, 1956). Erratics beyond the normally accepted glacial boundary shown by plus signs.

Linking the Old Kentucky River to the Teays

Because the downstream portion of the Old Kentucky River basin north of Cincinnati has been greatly modified by erosion and buried by glacial drift, its connection to the trunk Teays River valley in central Ohio is not obvious. This, plus the extensive modification of the Old Kentucky River southwest of Cincinnati along the modern Ohio River as far as Carrollton, Kentucky, has even generated disagreement about whether the river was originally tributary to the Teays River valley. The belief that the Old Kentucky River flowed northeastward to the Teays (e.g., Fowke, 1898, 1933; Malott, 1922; Wayne, 1952; Durrell, 1961) has been satisfactorily demonstrated over the years, even though its buried course north of Dayton still is not clearly delineated. Based on the gradual northward decline of the bedrock floor of the valley, including the projection of its gradient beneath glacial drift to the Teays River valley in central Ohio, there seems little doubt that the entrenched Old Kentucky River valley once flowed northward. Additional evidence from south of the glacial boundary supporting a link with the Teays River to the north include: (1) entrance angle of tributaries (Malott, 1922; Teller, 1973), (2) provenance of bedload deposits (Teller, 1973), (3) northward gradient on bedrock floor (Swadley, 1971), and (4) northward increase in meander wavelength (Dury and Teller, 1975).

The evolution of names for the major rivers in the Old Kentucky River system is discussed in Teller (1973). The names currently in use are shown in Figure 2.

MORPHOLOGY OF VALLEYS

Region north of Cincinnati

Northward from the southwestern corner of Ohio, under the cities of Hamilton, Middletown, Franklin, and Dayton, the buried Teays-age Old Kentucky River valley is more than a kilometer wide and closely approximates the course of the Great Miami River (Fig. 3). The ancient valley is traced only by plots of water well depths to bedrock (Norris, 1948; Cummins, 1959; Soller, 1986). The floor of this buried bedrock valley is below 150 m (500 ft) in elevation and continues the northward gradient at about 10 cm/km. Down-valley from Dayton, studies of thousands of water-well logs, and some test drilling, have not yet clearly delineated the lower reaches of the Old Kentucky River valley, where it must have continued northeast toward Springfield or possibly north toward Troy (see Cummins, 1959; Soller, 1986).

Region south of Cincinnati

Old Kentucky River valley. South of the glacial boundary (Fig. 1), many now-abandoned segments of valleys that were part of the Old Kentucky River system are topographically distinct. Because many major modern rivers, such as the Ohio, Great Miami, Licking, and Kentucky (Figs. 2, 3), have been established within or close to the valleys of the ancient rivers, erosion has been substantial, and their original outline and morphology cannot everywhere be clearly discerned.

The main Old Kentucky River channel is preserved as scattered but distinct sediment-veneered valley segments with flat to highly dissected floors (Fig. 4) from the western suburbs of Cincinnati along the Great Miami River valley, southwestward along the Ohio River to Carrollton, Kentucky, and then south along the modern Kentucky River past Frankfort (Figs. 1, 3). These valley remnants have been mapped on 1:24,000-scale topographic maps by the U.S. Geological Survey in its Geologic Quadrangle (GQ) Map series for the state of Kentucky, published during the 1960s and 1970s. In a series of reports, Jillson (1943; 1944a, b; 1945a; 1946a–d; 1947a, b; 1948a, b) described and mapped segments of the Old Kentucky River channels from south of Lexington to the Ohio River. In 1945, Jillson presented a summary of his views and compiled an annotated bibliography on this ancient river system (Jillson, 1945b). A general outline of the "old drainage systems" of the region, including that of the Old Kentucky River, was presented by Leverett (1929).

The reconstructed course of the Old Kentucky River valley follows a broadly meandering pattern that is entrenched 50 to 100 m below the upland level, and lies as much as 100 m above the modern trunk river. In many places the meandering pattern of modern valleys (Fig. 3) coincides with the reconstructed route of the Old Kentucky River. This ancient valley system has been recognized by the U.S. Geological Survey at least as far upstream as Lee County, 80 km southeast of Lexington, in the headwaters of the modern Kentucky River watershed, where it appears mainly as small (mappable) areas of fluvial sand and gravel, rather than as distinct valley remnants; topographically identifiable portions extend upstream past the Boonesborough fault zone southeast of Lexington. The reconstructed meander wavelength increases northward, reaching more than 7 km near Cincinnati.

The fragmentary nature of the Old Kentucky River valley, especially in the upstream portion, makes generalizations about its width and gradient difficult. South of the Ohio River at Carrollton, the abandoned valleys are mainly less than a half kilometer wide, whereas down valley toward Cincinnati, their width increases to nearly a kilometer. The bedrock floor of the valley declines along its strongly meandering course from an elevation of about 275 m (900 ft) in the Boonesborough fault zone south of Winchester, Kentucky, to about 205 m (670 ft) at Carrollton, and then to about 190 m (620 ft) near Cincinnati; the gradient downstream from Carrollton is 9.5 cm/km (Swadley, 1971).

Old Licking River valley. The Old Licking River drained an area that was nearly as large as the Old Kentucky River watershed south of Cincinnati. Figure 3 shows the modern areas drained by these two rivers, which are similar in size to their Teays-age ancestors. The Old Licking River joined the Old Kentucky River near Hamilton, Ohio (Fig. 2). Most of the downstream portion of this valley, north of Cincinnati, and most of the tributary Manchester River valley have been destroyed by erosion along modern and ancient courses of the Ohio River, which closely followed these old valleys.

The valley of the Old Licking River, also called the Claryville River valley (Teller and Last, 1981), is topographically distinct in many places over a distance of more than 150 km upvalley from its junction with the old Manchester valley. Although the valley floor and fluvial sediment cover have been severely eroded by the modern Licking River and its tributaries, many well-preserved segments lie 30 to 60 m above the modern rivers and 30 to 70 m below the uplands (Fig. 2). These remnants are shown clearly on U.S. Geological Survey Quadrangle Maps; Luft (1980) has described this valley and reconstructed its course as far south as Nicholas County. The average width of the valley increases northward, reaching 1 km near its mouth. Abandoned segments of the old South Fork River, the major tributary to the Old Licking, join near Falmouth, Kentucky (Fig. 2). This tributary and its deposits have been discussed by Luft (1986), who presented evidence that South Fork drained a much larger region before its headwaters were captured by the Old Kentucky River. According to Luft (1986), this helps explain the abrupt westward turn of the Old Kentucky River between Irvine and Lexington (see Fig. 3), the presence of valleys that are intermediate in elevation between subupland and Teays valleys, and the gravel clast lithology of the old South Fork valley remnants.

Luft (1980) plotted the longitudinal profile of the bedrock thalweg along the Old Licking River. Over a channel distance of 155 km, the elevation of the bed declines from about 220 m (720 ft) to 190 m (620 ft) at the junction with the Manchester River valley, for an average gradient along its meandering course of about 19 cm/km. This is very similar to gradients reported by Luft (1980) for the Teays valley in Ohio and West Virginia. The

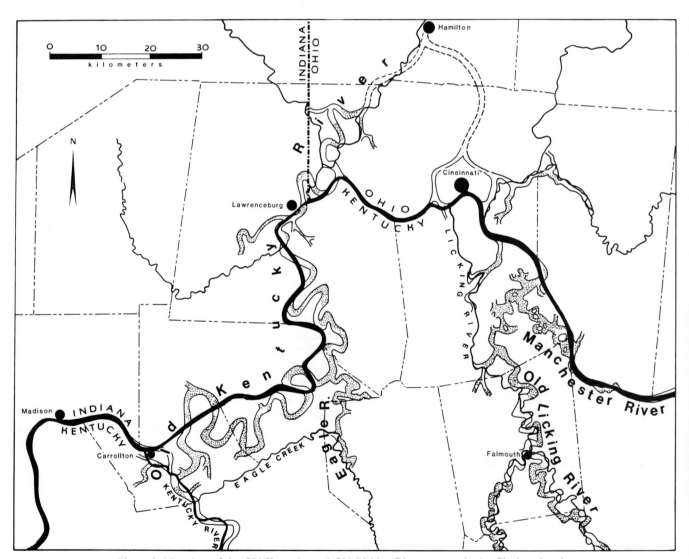

Figure 2. Meanders of the Old Kentucky and Old Licking River systems in the Cincinnati region, showing remaining segments (stippled) of the Teays-age valley floor (after Teller, 1973). Area south of Falmouth after Luft (1980).

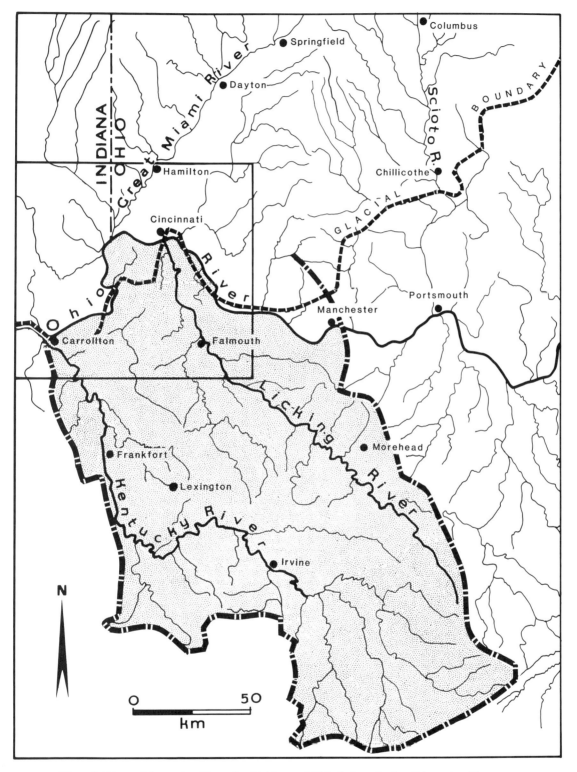

Figure 3. Rivers of the modern Kentucky, Licking, and Ohio drainage basins of the region. The present Kentucky and Licking River basins are indicated by shaded area. South of the glacial boundary (dashed line), the modern rivers developed close to or coincident with their ancestral courses; east of Manchester, valleys were once part of the Teays River basin.

Figure 4. View of abandoned Old Kentucky River valley near Ohio River at Rising Sun, Indiana.

modern Licking River, which closely follows the abandoned valley, has a gradient over the same distance of about 26 cm/km (Luft, 1980). Meanders along the Old Licking River channel have an average wavelength of about 3 km.

SEDIMENT WITHIN VALLEYS

Introduction

Sediment in the Old Kentucky River valley and its main tributaries is mainly fluvial sand and gravel. Closer to the glacial boundary, notably along the lower reaches of the Old Licking River valley, lacustrine clay and silt overlie the coarser sediment. Fluvial sediment is not confined to the floor of the old valleys, but also occurs between this level and the upland as terraces and "smears" on the valley walls. In addition, because many parts of the Old Kentucky River system have been entrenched, terraces and isolated fine to coarse deposits also lie below the floor of the old valleys. Our description is confined to those deposits that lie on the bedrock floor of the Old Kentucky and Old Licking River systems.

Old Kentucky River valley

In the Old Kentucky River valley itself, south of the glacial boundary, fluvial deposits on the bedrock floor vary from a thin veneer to a thickness of more than 16 m. Rarely is a clear depositional or erosional morphology, relating to the ancient river, still present in these valley remnants, although slip-off slopes on the inner bend of meanders exist in a few places. As with the underlying bedrock surface, the elevation of these deposits declines toward the north. Sediment along the length of the old river is deeply weathered and everywhere leached of carbonates, with even the bedrock below weathered in places.

Sediment characteristics are similar along the length of the Old Kentucky River, where they are described as Pliocene and Pleistocene "high-level fluvial deposits" on U.S. Geological Survey Quadrangle Maps. Swadley (1971), Jillson (1943; 1944a, b; 1945a, b; 1946a–d; 1947a, b; 1948a, b; 1963), Gooding and Wayne (1961, p. 129–130), Hester (1965), and others have also described these sediments in various places. Basically, the fluvial deposits consist of clayey silt, sand, and gravel, which are typically nonbedded and poorly sorted. Cobble sizes are commonly present. Gravel occurs throughout the sequence but is most abundant at or near the base. The lithology of the >2-mm-size fraction has been summarized by Swadley (1971, p. 130) as consisting of four types of material:

... quartz and chert pebbles, fragments of chert and silicified limestone, quartz geodes, and blocks of sandstone. The quartz and chert pebbles are white, yellow, or brown, well rounded, and generally less than 2 inches [5.1 cm] in diameter, resembling those in the Pennsylvanian-age conglomerate of southeastern Kentucky. Chert and silicified limestone occur as subangular to subrounded pebbles, cobbles, and blocks as much as 12 inches [30.5 cm] across. Most are some shade of brown, but others are gray and white; some include silicified corals and crinoids of probable Mississippian age. Geodes containing yellow or brown quartz are present throughout the deposits. They range from 1 to 18 inches [2.5 to 45.7 cm] in diameter and were probably derived from Mississippian-age limestone of south-central Kentucky. The least abundant component of the gravel, blocks of light-brown, medium- to coarse-grained micaceous sandstone as much as 2 feet [0.6 m] across, occur locally.

Deposits in the Old Kentucky River valley have been correlated with the Tertiary Irvine Formation and Lafayette gravels by Campbell (1898), Miller (1914, 1919), and McFarlan (1943). Miller (1919, p. 170) stated that the Lafayette gravels "have yielded fragmentary remains of mammals (such as tapir and an extinct species of deer)."

Unlike the Old Licking River system, southeast of Cincinnati, which was tributary to the Old Kentucky River, there is an almost complete absence of lacustrine clay and silt over the coarser fluvial sediment. Only a few localities of this rhythmically laminated clay and silt are known from within the Old Kentucky River valley, or in sub-upland depressions west of Cincinnati where, in places, it is calcareous and locally underlies till (Ettensohn, 1974; Ettensohn and Glass, 1978; Teller, 1970, p. 111, 112).

Old Licking River valley

Fluvial sand and gravel overlies the bedrock floor of the Old Licking River valley along the lower few hundred kilometers of its abandoned channel, and is delineated on U.S. Geological Survey Quadrangle Maps as "high-level fluvial deposits" of Pliocene(?) and Pleistocene age. Luft (1980) presented a map of the ancient Licking River system, showing the extent of its deposits, and later (1986) did the same for the South Fork of this old river. In places, notably in the southern half of the drainage basin, upland gravels are included with these valley deposits on USGS maps. Other deposits may be related, such as the Irvine Formation and Lafayette gravels (e.g., Campbell, 1898; Miller, 1895, 1919; McFarlan, 1943), which overlie bedrock at upland and sub-upland elevations over much of central Kentucky, as well as deeply weathered upland sand near the mouth of the Old Licking River (Leverett, 1929; Teller, 1973; Luft, 1980; and references therein).

Fluvial sediment within the Old Licking River valley and its main tributary, South Fork, is thickest in the downstream area, reaching 22 m near Falmouth, and thinning to less than 10 m, 80 km up-valley from there (Luft, 1980, 1986). According to Luft (1980), these deposits consist of, in decreasing abundance, silt, clay, gravel, and sand that is deeply weathered and either crudely bedded or lacking recognizable bedding. Gravel, commonly in a sandy to clayey matrix, is mainly concentrated near the base of the deposit and is as thick as 2 m in the channel thalweg; the gravel "generally grades upward into an ill-defined gravelly sand and gravel zone and then to clayey silt with sparse pebbles. . . . No fossils have been found in Teays-age Licking deposits" (Luft, 1980, p. 3–4).

The sand fraction, which locally is moderately well sorted and subangular to angular, is 99 percent quartz; its most likely source is the Pennsylvanian Lee Formation (Luft, 1980). The lithology of the gravel fraction resembles that in the Old Kentucky River valley, being derived mainly from Mississippian and Pennsylvanian rocks and from a few local, lower Paleozoic sources. In decreasing order of abundance, Luft (1980) has summarized the pebble, cobble, and boulder lithology as consisting of dark reddish brown jasperoid chert as much as 30 cm in diameter from the Mississippian Newman Limestone, well rounded to subrounded quartz pebbles derived from the Lee Formation, rounded pebbles and cobbles of Lee Formation sandstone and conglomerate, yellowish brown fossiliferous chert from the Silurian Brassfield Formation, and pebble- to cobble-sized fragments of siltstone, sandstone, silicified limestone, and chert from Ordovician- to Mississippian-aged rocks. In addition, there are occasional rounded coal pebbles within deposits of the Old Licking River (Hays, 1951; Durrell, 1961; Luft, 1980), which were derived from the Pennsylvanian Breathitt or Lee Formations. The presence of coal and the absence of geodes distinguish these fluvial deposits from those in the Old Kentucky River valley.

In the South Fork of the Old Licking River valley, which joins the main Old Licking at Falmouth (Fig. 2), gravels are similar in composition to those of the Old Kentucky River valley and contain no coal (Luft, 1986).

The Manchester River valley (Fig. 2), which heads near Manchester, Ohio, 80 km to the southeast of Cincinnati (Fig. 1), has been nearly destroyed by development of the modern Ohio River. Except for a few remnants near its junction with the Old Licking River, no deposits related to this ancient river have been identified. Gibbons and others (1975) described sediment from the Manchester River valley as sand, containing gravel composed almost entirely of angular to subangular fragments of brown chert and pebbles of pale yellow silicified limestone or dolomite, with a few rounded quartz pebbles.

Overlying the fluvial sediment in the Old Licking River valley, almost as far upstream as the junction with South Fork at Falmouth (Fig. 2), is the Claryville Clay. This sediment, as much as 15 m thick, is mainly a finely laminated lacustrine sandy to silty clay, with a general increase in grain size with depth (Teller and Last, 1981). The unit grades southward into fluvial deposits, and may be equivalent to the upper part of the fine-grained fluvial deposits south of Falmouth. "Tributary stream sediments and possible deltaic deposits are locally present within some arms of the former lake" (Luft, 1980, p. 5). The sediment has been deeply weathered and eroded, although topography on the Claryville Clay is distinctly smoother than on the adjacent bedrock.

On the basis of variations in clay mineralogy, grain size, and sedimentary structures, Teller and Last (1981) recognized three lithologic units in the Claryville Clay, which they correlated with the three units identified by Ettensohn and Glass (1978) at other ancient ponded sediment localities in the region. Coal grains have been noted in these clays (Martin, 1977), and about 5 percent of the microfossil assemblage in one core is interpreted as having been reworked from Carboniferous rocks (Teller and Last, 1981).

HISTORY OF VALLEY DEVELOPMENT

Earliest Tertiary drainage

The history of Tertiary drainage in the Interior Low Plateau and adjacent Appalachian Plateau, south of the glacial boundary, has long been discussed. Deposits of deeply weathered sandy gravel, containing distinctive resistant components such as quartz and quartzite pebbles, chert, and geodes, have been described throughout much of Kentucky at upland and subupland levels. Most have concluded that these sediments were deposited by graded rivers at or near baselevel. The age of these gravels has generally been considered Tertiary, and their deposition has been related to the "cyclic" history of erosion in the region, mainly to the Lexington–Highland Rim peneplain and to the subsequent Parker strath entrenchment of rivers. Thornbury (1965, p. 193) stated that these deposits, the Lafayette gravel and the equivalent Irvine Formation, ". . . represent a lag concentrate of insoluble siliceous materials that accumulated under mid- to late-Tertiary subtropical climatic conditions on the Lexington–Highland Rim peneplain. Uplift of this erosion surface was followed by cutting of the Parker strath, and during its development, large quantities of gravel were swept off the peneplain surface and concentrated on the Parker strath along such rivers as the Ohio, Tennessee, Cumberland, and Kentucky." The Teays system, including the Old Kentucky and Old Licking Rivers, presumably was developed during the Parker subcycle of erosion.

We consider the sandy gravel lag at or near upland levels to have been deposited either prior to the development of the Teays River system or during an early, pre-entrenchment phase. The scattered remnants of these deposits, even as mapped on

1:24,000-scale U.S. Geological Survey Quadrangle Maps, do not allow a reconstruction of this ancient river system. However, the concentration of such upland deposits close to the entrenched (now abandoned) valleys of the Old Kentucky and Old Licking Rivers, suggests a relation between them. It seems likely that ancient Tertiary drainageways, meandering over the Lexington–Highland Rim peneplain (Fig. 5A), were entrenched as baselevel declined, to establish the valleys of the Old Kentucky and Old Licking Rivers and their tributaries (Fig. 5B).

Establishment of the Teays drainage system

Tight (1903), in the original detailed discussion of the Teays valleys in West Virginia and adjacent areas, ascribed a Tertiary age to the system. The initial entrenchment of the meandering rivers below the flat Tertiary uplands may have come about as a result of early glacier buildup in Antarctica that produced a fall in sea level. The establishment of permanent ice on the Antarctic continent probably occurred during late Miocene time (e.g., Kennett, 1977, 1982; Kemp, 1978; Adams and others, 1977; Schnitker, 1980), although river entrenchment in the interior of North America probably would not have occurred until long after this. Other major global sea-level changes that may have affected valleys of the Teays system occurred during the late Pliocene, 2.5 to 3.0 Ma, when there was substantial climatic cooling and ice cap growth (e.g., Cronin, 1988; Kennett, 1982; Ruddiman and Raymo, 1988), and in the Pleistocene after 0.9 Ma, when there was a shift toward higher amplitude climatic fluctuations (e.g., Ruddiman and Raymo, 1988).

The relation of meander wavelength to drainage-basin size of the Old Kentucky River system led Dury and Teller (1975) to conclude that the initiation and growth of these valleys required a long, but not necessarily continuous, period of high discharge that exceeded that in the region today. The failure of these rivers to substantially broaden their valleys after entrenchment suggests that a new constant baselevel was never established for any great length of time. However, some valley expansion did occur, and numerous meander loops were cut off (Fig. 5C).

Because the elevation of the bedrock floor of the Old Kentucky River valley gradually declines northward to where it merges with the buried Teays River valley of central Ohio (Fig. 1), it seems likely that the events controlling entrenchment were the same throughout the Teays system. Only events on a continental scale could have affected such a large drainage system.

After having entrenched their channels more than 30 m below the upland levels, these Teays tributaries began to aggrade, infilling valleys and cut-off meander loops (Fig. 5D) with sand and gravel that contains the same resistant fraction as lay on the floors of their upland predecessors. Because of their similarity, all of these deposits are commonly mapped as the same sedimentary unit (e.g., Irvine Formation, Lafayette gravels, "high-level fluvial deposits"), even though many of those along the Old Kentucky River system lie in discernible valley remnants at elevations well below the upland.

The partial infilling of the entrenched Old Kentucky River valley system probably came about as a result of a reduction in the ratio between river discharge and sediment load. If Luft's (1986) interpretation that the Old Kentucky River captured the headwaters of the South Fork of the Old Licking River is correct, then this may have played a role. A rise in baselevel to the north, possibly in conjunction with glacial damming over the lower reaches of the system, could have been involved in producing the fining-upward sequence in the valleys.

Thus, the meandering pattern of the Old Kentucky River system is relict from the sub-upland phase of Tertiary drainage (Fig. 5A). During entrenchment (Fig. 5B), the discharge of these rivers may have remained constant, or even have been greater, but subsequent aggradation in the valleys required a reduced discharge/sediment ratio. The rivers that were eventually established on the aggraded floors of these valleys probably meandered, not only within the main preaggradation bedrock channel (Fig. 5D), but also within the previously abandoned cut-off meander loops.

Changes in the Old Kentucky River system—An introduction

Major modifications to the Old Kentucky River drainage system followed the Tertiary entrenchment and aggradation phases. All changes did not come about at the same time, nor for the same reason. In part, the mechanism inducing change was glaciation, which repeatedly invaded the region as far south as Cincinnati. Glaciation not only impounded waters in the Old Kentucky River drainage basin, but also forced the rivers in the headwaters of the Teays River itself, upstream from the junction with the Old Kentucky River, to overflow southward around the ice margin into the Manchester River valley of the Old Kentucky River system (Figs. 1, 2). Headward growth of the Old Ohio River valley across the ancient divide at Madison, Indiana (Fig. 1), possibly combined with overflow resulting from glacial damming of the lower reaches of the Old Kentucky River, brought about the integration of the Old Ohio and Old Kentucky systems.

The modern drainageways of the region are parallel to, or coincide with, most of the old Tertiary river valleys (Fig. 3), although near the glacial border along the Ohio River their original size and shape has been largely destroyed. To the north, the Old Kentucky River system has been modified by multiple events of glacial and meltwater erosion, and now lies buried beneath thick glacial drift.

In the following sections we attempt to interpret the sequence of events leading to abandonment and entrenchment of valleys in the Old Kentucky River system and to the establishment of modern rivers. The interrelations of the causal events—namely, piracy, glacial damming, overflow from Teays River headwaters—are not fully understood, especially in terms of the number of glacial events leading to the changes. Even the exact sequence of some changes cannot be determined with certainty. There is, however, enough evidence to establish, with confidence, a general history of the evolution of drainage in the region.

Piracy by the Old Ohio River

Further entrenchment of the Teays-age Old Kentucky River valley below its already entrenched and aggraded floor (Fig. 5D) took place after its waters were integrated into the Old Ohio River (Fig. 1). The southwestward-draining Old Ohio River originally headed along the Silurian escarpment, which served as the divide between the Ohio watershed and that of the northward-draining Old Kentucky River system (Fig. 1). Once the divide near Madison, Indiana, only 20 km west of the Old Kentucky River at Carrollton, was breached by the Old Ohio River, it was only a matter of time before the river upstream from Carrollton (Fig. 3) was captured. Entrenchment of the aggraded valley floor followed. Eventually the Old Kentucky River downstream (northeast) of Carrollton was integrated into the expanding Ohio River system. In places, a straighter, more direct route developed, and between Carrollton and Cincinnati the new Ohio River was established in a channel that only locally coincided with the ancestral valley (Fig. 2). Those meander loops along the Old Kentucky River south of Carrollton, which had been cut off prior to the aggradation phase (Fig. 5C), were again abandoned in favor of a steeper, straighter, and more direct route. It is mainly

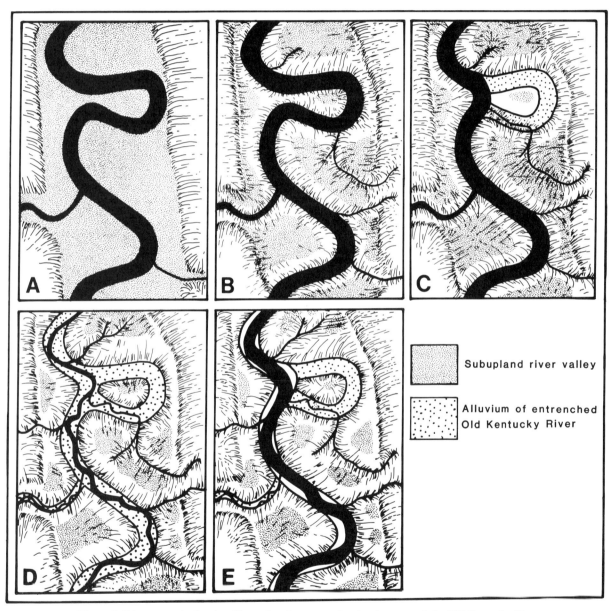

Figure 5. Diagrammatic sketches illustrating the progressive development of the Old Kentucky River from its sub-upland phase to its final entrenchment and abandonment. A, Meandering Tertiary river at sub-upland level. B, Entrenchment of meanders after fall in baselevel. C, Cutoff of meander loop after entrenchment. D, Aggradation of entrenched valley floor by underfit river. E, Entrenchment through alluvial fill and bedrock floor of Old Kentucky River after addition to Ohio River system.

those cut-off meanders that today remain as remnants of the Teays-age Old Kentucky River.

Although some have maintained that westward overflow into the Old Ohio River valley at Madison did not occur until after the downstream area of the Teays was dammed by glaciation (e.g., Swadley, 1971), Teller (1973) has outlined the reasons for believing that this occurred prior to glacial impoundment. The absence of fine-grained ponded sediment in the abandoned valleys of the Old Kentucky River, at least upstream from Lawrenceburg, Indiana (just west of Cincinnati), in contrast to the thick lacustrine sediment in the Old Licking River valley, is the major reason for believing that preglacial piracy, rather than erosion by proglacial lake overflow, caused the Old Kentucky River to be added to the headwaters of the Old Ohio River. In other words, had the upstream portions of the Old Kentucky River not been captured prior to glacial damming in central Ohio, water would have been impounded in that valley and lacustrine sediment would have been deposited on its floor—but no such sediment has been found. S. M. Totten (personal communication, 1989) believes that the flat, largely undissected topography near the Ohio River at Madison, together with the presence of shallow upland valleys in the Old Ohio basin of that area, indicates that rapid downcutting by glacially ponded waters, not piracy, was responsible for the integration of the Old Kentucky and Old Ohio River systems.

Southeast of Cincinnati, in the Old Licking River system (Fig. 2), lacustrine sediment does occur above the fluvial deposits. If this part of the Old Kentucky River drainage system had not yet been integrated into the expanding Ohio River basin, impoundment of water in the Old Licking River valley would have occurred as soon as ice overrode the lower reaches of the valley system west or south of Springfield, Ohio. If the Old Licking and Manchester Rivers had been added to the new Ohio River system (Fig. 6) prior to glaciation, impoundment would not have taken place until the ice advanced south of Hamilton, Ohio. Therefore, the presence of lacustrine deposits in the Old Licking valley cannot be used as evidence either for or against its integration with the new Ohio River system prior to glaciation, because in either situation water would have been ice-dammed in the valley at some stage.

The three distinct lacustrine units identified in the Claryville Clay of the lower Old Licking River valley (Teller and Last, 1981) and elsewhere in the region (Ettensohn and Glass, 1978) reflect three stages of glacial ponding in the valley remnants. It is not known if piracy had established the Old Licking River in a new, more deeply entrenched valley, along the lines of its modern route, prior to this glacial impoundment, but no lacustrine sediment is found along these younger valleys. As with the Old Kentucky River valley to the west, a straighter and steeper route on the aggraded floor of the old valley was developed by the younger underfit river, leaving the meander loops as abandoned remnants.

Establishment of the New Ohio River System

Glaciation was important in determining the final course of the new Ohio River, regardless of whether or not it was involved in the process of reversing the flow of the Old Kentucky River between Cincinnati and Carrollton and integrating it, along with the Old Licking River, with the Old Ohio River basin. Glaciation was responsible for severing the last link between the Old Kentucky River and Teays River watersheds. Once the lower valley of the Old Kentucky River was dammed by ice, all runoff from its drainage basin was forced to overflow westward through the Ohio River valley, and any remaining segment of the Old Kentucky River that still flowed northeastward between Carrollton and Hamilton (Fig. 2) would have reversed its course. Waters from the Old Licking and Manchester River basins, as well as from glacial melting, would have been added to this new Ohio River system, helping to establish and entrench a course along the lines of the present-day river (Leverett, 1929; Fowke, 1933; Wayne, 1952). In many places, both east and west of Cincinnati, the exact course of the new Ohio river probably was determined partly by the margin of the ice, as well as by the location of the aggraded floor of the old rivers.

Glaciation also dammed the headwaters of the Teays River system of southeastern Ohio (Fig. 1), causing overflow into the Old Kentucky River watershed at the divide near Manchester, Ohio (Fig. 3) (Tight, 1903; Leverett, 1902, 1929; Fowke, 1933). Tight (1903) noted that the entrance angles of tributaries to the Ohio River east of Manchester are barbed, and from this observation reconstructed the preglacial drainage for the entire headwater region of the Teays beyond the glacial boundary. Thus, not only did glaciation help establish the new Ohio River along its present course from Madison, Indiana, to Manchester, Ohio, but it also brought about the eventual integration of the headwaters of the Teays River into the new Ohio River basin. This latter event, however, may not have resulted from the same glaciation that was responsible for severing the link between the Old Kentucky and Teays River.

The age of the glacial event (or events) responsible for establishing the new Ohio River from the Old Kentucky River, Old Ohio River, and Teays River headwaters is not known with certainty. We believe that some of this integration of basins occurred by piracy prior to glaciation. It is possible that the widely scattered erratics south of the presently recognized glacial boundary, on the uplands of Kentucky (Fig. 1) (Jillson, 1924a, b, 1925, 1963; Leverett, 1929) and southeastern Ohio (Patton and Hicks, 1925; Leverett, 1929, p. 50; Ireland, 1943; Merrill, 1953), may represent the earliest glaciation, and that could have been the event that resulted in the demise of the Teays River system. In southeastern Ohio (Bonnett and others, this volume) and eastern Indiana (Bleuer, 1976; this volume), ponded sediment related to one or more early glacial episodes was deposited during a "reversed" magnetic polarity epoch, and therefore is more than

700,000 yr old. Measurements of remnant magnetism in the Claryville Clay of the Old Licking River valley yielded "normal" values (Teller and Last, 1981), suggesting an age date of <0.7 Ma (Brunhes Paleomagnetic Epoch), 0.8 to 0.9 Ma (Jaramillo event), or >1.6 Ma (Gilsa or older "normal" events). It is not known how these ponded sediments relate to the old glacial drift of the region or to the establishment of the modern Ohio River.

Deeply weathered, pre-Illinoian till mantles the uplands just south of the Ohio River in the Old Kentucky River watershed (Flint and others, 1959; Durrell, 1961; Swadley, 1971, 1979; Teller, 1972) and also is found in the main Old Kentucky River valley east of Carrollton (Swadley, 1971; Teller, 1973) and even below that level (Teller, 1973). Four pre-Illinoian tills have been identified in the region (Teller, 1972), with several workers proposing a "Nebraskan" age for the oldest (e.g., Swadley, 1979; Norton and others, 1983). These tills may represent glaciations that occurred long after the Old Kentucky River and Teays River watersheds had established separate identities. Undoubtedly, however, early Pleistocene, or perhaps pre-Pleistocene, glaciation played a major role in the demise of the ancient Teays River system.

Once established, the new Ohio River was eventually entrenched about 75 m below the floor of the Teays-age valleys, to a depth of more than 30 m below the modern flood plain of the river (Teller, 1973). This entrenchment is referred to as Deep Stage drainage, and developed sometime prior to Illinoian glaciation (Leverett, 1902, 1929; Ver Steeg, 1934; Rich, 1956; Wayne, 1956; Durrell, 1961). Deep Stage entrenchment occurred partly as a result of capture by the Old Ohio River watershed, as well as from multiple episodes of pre-Illinoian meltwater discharge.

Development of the modern course of the Ohio River at Cincinnati probably occurred in stages, beginning with abandonment of the northern loop through Hamilton (Fig. 6) during one of the pre-Illinoian glaciations. The river probably assumed the present course when Illinoian glaciers invaded the Cincinnati area (Teller, 1973).

SUMMARY

The Old Kentucky River drainage basin headed in southeastern Kentucky and joined the Teays River basin in west-central Ohio. The main rivers, the Old Kentucky and Old Licking, evolved from northward-flowing Tertiary rivers that meandered across the Lexington–Highland Rim peneplain, probably during a time when discharge was greater than in today's rivers of the region. Entrenchment, perhaps as a result of sea-level lowering by Tertiary glacial buildup in Antarctica, deepened these channels more than 30 m below their original sub-upland floors. Following a period during which some meander loops were cut off, the floors of these entrenched valleys were aggraded. This phase resulted from a reduction in the river discharge/sediment ratio, which may have come about because of a decline in precipitation; a rise in baselevel, related to glaciation in the Teays basin, may have been involved.

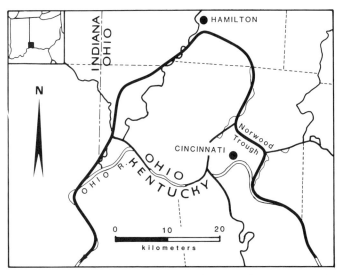

Figure 6. Initial phase of the new Ohio River, showing its route north around Cincinnati via Hamilton (after Teller, 1973).

The fining-upward sequence of fluvial gravel, sand, and silt in the remnants of the now-abandoned Old Kentucky River system is composed of a deeply weathered but distinctive gravel lithology, including quartz, chert, silicified limestone, geodes, and coal, derived mainly from late Paleozoic rocks of southeastern Kentucky. Entrenchment of the Teays-age Old Kentucky River system came about after the divide near Madison, Indiana, only a few kilometers to the west, was breached, leading to the capture of the entire Old Kentucky River system south of Carrollton, Kentucky, by the Old Ohio River system. Only the scattered, previously aggraded, cut-off meanders of the old system were preserved after this entrenchment.

Glacial damming of the Old Kentucky River system north of Cincinnati brought about the complete integration of the Old Kentucky and Old Ohio River drainage basins, and further entrenched the new drainageways. The link between the Teays and Old Kentucky River systems came to an end at this time. Glacial damming also caused overflow from the headwaters of the Teays River basin, across the divide at Manchester, Ohio, expanding the new Ohio River farther eastward into West Virginia, southeastern Ohio, and eastern Kentucky.

The age of the entrenched Old Kentucky River system is Tertiary. Changes to its network of valleys came about in stages, probably beginning by piracy during the Tertiary and continuing until early glaciation brought about its final abandonment.

ACKNOWLEDGMENTS

We thank Anne Flynn for typing, Ron Pryhitko for drafting, and Dave Mickelson and Stan Totten for helpful reviews of this paper.

REFERENCES CITED

Adams, C., Benson, R., Kidd, R., Ryan, W., and Wright, R., 1977, The Messinian salinity crisis and evidence of late Miocene eustatic changes in the world ocean: Nature, v. 269, p. 383–386.

Bleuer, N. K., 1976, Remnant magnetism of Pleistocene sediments of Indiana: Indiana Academy Science Proceedings, v. 85, p. 277–294.

Burger, A. M., Keller, S. J., and Wayne, W. J., 1966, Map showing bedrock topography of northern Indiana: Indiana Geological Survey Miscellaneous Map 12, scale 1:500,000.

Campbell, M. R., 1898, Description of the Richmond Quadrangle: U.S. Geological Survey Atlas Folio 46, 4 p.

Cronin, T. M., 1988, Evolution of marine climates of the U.S. Atlantic coast during the past four million years: Philosophical Transactions Royal Society London B, v. 318, p. 661–678.

Cummins, J. W., 1959, Probable surface of bedrock underlying the glaciated area in Ohio: Ohio Water Plan Inventory Report 10.

Durrell, R. H., 1961, The Pleistocene geology of the Cincinnati area, *in* Guidebook for Field Trips, Annual Meeting, Cincinnati: Geological Society of America, p. 45–57.

Dury, G. H., and Teller, J. T., 1975, Paleogeomorphic and paleoclimatic implications of "preglacial" meanders in the Cincinnati area: Geology, v. 3, p. 585–586.

Ettensohn, F. R., 1974, The pre-Illinoian lake clays of the Cincinnati region: Ohio Journal of Science, v. 74, p. 214–226.

Ettensohn, F. R., and Glass, H. D., 1978, Clay-mineral stratigraphy of pre-Illinoian lake clays from the Cincinnati region: Journal of Geology, v. 86, p. 393–402.

Flint, R., Colton, R., Goldthwait, R., and Willman, H., 1959, Glacial map of the United States east of the Rocky Mountains: Geological Society of America, scale 1:1,750,000.

Fowke, G., 1898, Preglacial drainage in the vicinity of Cincinnati; Its relation to the origin of the modern Ohio River, and its bearing upon the question of the southern limits of the ice sheet: Denison University Science Laboratory Bulletin, v. 11, p. 1–10.

—— , 1933, The evolution of the Ohio River: Indianapolis, Indiana, Hollenback Press, 273 p.

Gibbons, A. B., Kohut, J. J., and Weiss, M. P., 1975, Geological map of the New Richmond Quadrangle, Kentucky-Ohio: U.S. Geological Survey Geologic Quadrangle Map GQ-1228, scale 1:24,000.

Gooding, A. M., and Wayne, W. J., leaders, 1961, Road log of second day, *in* Guidebook for Field Trips, Cincinnati Meeting: Geological Society of America, p. 107–130.

Hays, F. R., 1951, The Pleistocene history of the Cincinnati area: Compass, v. 28, p. 131–139.

Hester, N. C., 1965, A study of high level valleys in southwest Hamilton County, Ohio [M.S. thesis]: Cincinnati, Ohio, University of Cincinnati, 72 p.

Ireland, H. A., 1943, Pre-Illinoian glaciation in southeastern Ohio: Ohio Journal of Science, v. 43, p. 180–181.

Jillson, W. R., 1924a, Glacial pebbles in eastern Kentucky: Science, v. 60, no. 1544, p. 101–102; reprinted in Kentucky Geological Survey, 1927, ser. VI, v. 30, p. 123–126.

—— , 1924b, Glaciation in eastern Kentucky: Pan American Geologist, v. 42, p. 125–132; reprinted in Kentucky Geological Survey, 1927, ser. VI, v. 30, p. 127–135.

—— , 1925, Early glaciation in Kentucky: Pan American Geologist, v. 44, p. 17–20; reprinted in Kentucky Geological Survey, 1927, ser. VI, v. 30, p. 137–141.

—— , 1943, An abandoned Pliocene channel of the Kentucky River: Frankfort, Kentucky, Roberts Printing Co., 16 p.

—— , 1944a, The Pot Ripple abandoned channel of the Kentucky River: Frankfort, Kentucky, Roberts Printing Co., 30 p.

—— , 1944b, The Elkhorn abandoned channel of the Kentucky River: Frankfort, Kentucky, Roberts Printing Co., 34 p.

—— , 1945a, The Drennon abandoned channel of the Kentucky River: Frankfort, Kentucky, Roberts Printing Co., 26 p.

—— , 1945b, The Kentucky River; An outline of the drainage modification of a master stream during geologic time: Frankfort, Kentucky, The State Journal, 104 p.

—— , 1946a, The Easterday abandoned channel of the Ohio River: Frankfort, Kentucky, Roberts Printing Co., 43 p.

—— , 1946b, The English abandoned channel of the Kentucky River: Frankfort, Kentucky, Roberts Printing Co., 35 p.

—— , 1946c, The Providence abandoned channel of the Kentucky River: Frankfort, Kentucky, Roberts Printing Co., 19 p.

—— , 1946d, The Nonesuch abandoned channel of the Kentucky River: Frankfort, Kentucky, Roberts Printing Co., 25 p.

—— , 1947a, The Warwick abandoned channel of the Kentucky River: Frankfort, Kentucky, Roberts Printing Co., 38 p.

—— 1947b, Pliocene deposits of the lower Kentucky River valley: Frankfort, Kentucky, Roberts Printing Co., 15 p.

—— , 1948a, The Pleasant Hill abandoned channel of the Kentucky River: Frankfort, Kentucky, Roberts Printing Co., 36 p.

—— , 1948b, The Hickman abandoned channel of the Kentucky River: Frankfort, Kentucky, Roberts Printing Co., 29 p.

—— , 1963, Delineation of the Mesozoic course of the Kentucky River across the inner Bluegrass Region of the state: Frankfort, Kentucky, Roberts Printing Co., 24 p.

Kemp, E. M., 1978, Tertiary climatic evolution and vegetation history in the southeast Indian Ocean region: Palaeogeography, Palaeoclimatology, Palaeoecology, v. 24, p. 169–208.

Kennett, J. P., 1977, Cenozoic evolution of Antarctic glaciation, the circum-Antarctic Ocean, and their impact on global paleoceanography: Journal of Geophysical Research, v. 82, p. 3843–3860.

—— , 1982, Marine geology: Englewood Cliffs, New Jersey, Prentice-Hall, 813 p.

Leverett, F., 1902, Glacial formations and drainage features of the Erie and Ohio basins: U.S. Geological Survey Monograph 41, 802 p.

—— , 1929, The Pleistocene of northern Kentucky: Kentucky Geological Survey, ser. VI, v. 31, p. 1–80.

Luft, S., 1980, Map showing preglacial and pre-Illinoian deposits of the Licking River, north central Kentucky: U.S. Geological Survey, Miscellaneous Field Studies Map MF-1194, 13 p.

—— , 1986, The South Fork of the Licking River; Eastern Kentuck's major late Tertiary river?: Southeastern Geology, v. 26, no. 4, p. 239–249.

Mallot, C. A., 1922, The physiography of Indiana, *in* Handbook of Indiana geology: Indiana Department of Conservation, Division of Geology Publication 21, pt. 2, p. 132–138.

Martin, R., 1977, Structures in clayey sediments from a pre-Illinoan glacial lake, Claryville Area, Campbell Co., Kentucky [M.S. thesis]: Cincinnati, Ohio, University of Cincinnati, 124 p.

McFarlan, A. C., 1943, Geology of Kentucky: Lexington, University of Kentucky, 531 p.

Merrill, W. M., 1953, Pleistocene history of part of the Hocking River valley, Ohio: Ohio Journal of Science, v. 53, p. 143–158.

Miller, A. M., 1895, High level gravel and loam deposits of Kentucky Rivers: American Geologist, v. 16, p. 281–287.

—— , 1914, Geology of Franklin County: Kentucky Geological Survey, ser. IV, pt. 3, 87 p.

—— , 1919, The geology of Kentucky: Department of Geology and Forestry of Kentucky, ser. V, Bulletin 2, p. 168–174.

Norris, S. E., 1948, The bedrock surface and former drainage systems of Montgomery County Ohio: Ohio Journal of Science, v. 48, no. 4, p. 146–150.

Norton, L. D., Hall, G. F., and Goldthwait, R. P., 1983, Pedologic evidence of two major pre-Illinoian glaciations near Cleves, Ohio: Ohio Journal of Science, v. 83, p. 168–176.

Patton, L. T., and Hicks, C., 1925, Notes on the occurrence of glacial material beyond the border of the drift in Muskingum County, Ohio: Ohio Journal of Science, v. 25, p. 97–98.

Rich, J. L., 1956, Pre-Illinoian age of upland till in southeastern Indiana, southwestern Ohio, and adjacent parts of Kentucky: Geological Society of America Bulletin, v. 67, p. 1756.

Ruddiman, W. F., and Raymo, M. E., 1988, Northern Hemisphere climate regimes during the past 3 Ma; Possible tectonic connections, *in* Shackleton, N., West, R., and Bowen, D., eds., The past three million years; Evolution of climatic variability in the North Atlantic region: Philosophical Transactions of the Royal Society of London, B. Biological Sciences, v. 318, p. 411–430.

Schnitker, D., 1980, Global paleoceanography and its deep water linkage to the Antarctic glaciation: Earth Science Reviews, v. 16, p. 1–20.

Soller, D. R., 1986, Preliminary map showing the thickness of glacial deposits in Ohio: U.S. Geological Survey Miscellaneous Field Studies Map MF-1862, scale 1:500,000.

Swadley, W. C., 1971, The preglacial Kentucky River of northern Kentucky: U.S. Geological Survey Professional Paper 750-D, p. D127–D131.

——, 1979, New evidence supporting Nebraskan age for origin of Ohio River in north-central Kentucky: U.S. Geological Survey Professional Paper 1126-H, p. H1–H7.

Teller, J. T., 1970, Early Pleistocene glaciation and drainage in southwestern Ohio, southeastern Indiana, and northern Kentucky [Ph.D. thesis]: Cincinnati, Ohio, University of Cincinnati, 115 p.

——, 1972, Significant multiple pre-Illinoian till exposure in southeastern Indiana: Geological Society of America Bulletin, v. 83, p. 2181–2188.

——, 1973, Preglacial (Teays) and early glacial drainage in the Cincinnati area, Ohio, Kentucky, and Indiana: Geological Society of America Bulletin, v. 84, p. 3677–3788.

Teller, J. T., and Last, W., 1981, The Claryville Clay and early glacial drainage in the Cincinnati, Ohio, region: Palaeogeography, Palaeoclimatology, Palaeoecology, v. 33, p. 347–378.

Thornbury, W. D., 1965, Regional geomorphology of the United States: New York, John Wiley & Sons, 609 p.

Tight, W. G., 1903, Drainage modifications in southeastern Ohio and adjacent parts of West Virginia and Kentucky: U.S. Geological Survey Professional Paper 13, 111 p.

Ver Steeg, K., 1934, The buried topography of north-central Ohio and its origin: Journal of Geology, v. 42, p. 602–620.

Wayne, W. J., 1952, Pleistocene evolution of the Ohio and Wabash valleys: Journal of Geology, v. 60, p. 575–585.

——, 1956, Thickness of drift and bedrock physiography of Indiana north of the Wisconsin glacial boundary: Indiana Geological Survey Report of Progress, v. 7, 70 p.

MANUSCRIPT ACCEPTED BY THE SOCIETY JUNE 29, 1990

Printed in U.S.A.

Origin and history of the Teays drainage system; The view from midstream

Henry H. Gray
Geological Survey Division, Indiana Department of Natural Resources, 611 North Walnut Grove Avenue, Bloomington, Indiana 47405

ABSTRACT

Configuration of the buried part of the Teays Valley system across western Ohio, Indiana, and eastern Illinois suggests that the Teays is not a preglacial system, but rather, that it was formed marginal to a major glacier earlier than that which created the Ohio River, probably in similar fashion by consolidating and diverting fragments of older drainages. Pertinent criteria include (1) the relatively straight gorge that (2) crosses at least three regionally high areas with (3) few tributaries that join at grade. Also significant are (4) the depth of the gorge across a broad limestone plateau that has (5) relatively shallow karst development. These features imply a youthful valley system that was destroyed by burial before reaching a mature stage.

Although thousands of drillholes and seismic datum points in Indiana alone detail the bedrock surface and the nature of the unconsolidated deposits that overlie it, many questions remain regarding the evolution of that surface. Fluvial and lacustrine deposits associated with the earliest presently known till in Indiana (>0.8 Ma) fill eastern parts of the Teays gorge. Are there tills of pre-Teays age, and is any part of a pre-Teays valley system identifiable? What stratigraphic criteria distinguish those parts of the Teays that have been reoccupied and incorporated into younger, but now also buried, valley systems? The Blue River Strath of southern Indiana shares many characteristics with the type Teays and the ancestral Kentucky River valley and may be coeval with them; are there other such features? A regional approach to these and related questions should yield results.

INTRODUCTION

In west-central Ohio and east-central Indiana, a bedrock valley deeply buried by unconsolidated deposits and without surface expression was known by well drillers more than 100 years ago; in southeastern Ohio and adjacent parts of West Virginia, a group of high-level abandoned surface valleys were noted even earlier. The latter were first synoptically viewed by Tight (1903); 40 years later, Stout and others (1944) were among the first to visualize the buried valley system and to integrate it with its surface counterpart, by then long called the Teays Valley. In Indiana, Fidlar (1948), McGrain (1950), and Wayne (1956) extended the Teays Valley from Ohio and recognized its connection with the Mahomet Valley of Illinois, which had been delineated by Horberg (1945, 1950).

Since these studies of 30 years ago and more, many wells have been drilled, many seismic, gravity, and resistivity data have been taken, and much has been learned of the glacial stratigraphy and glacial history of this area. It is time for a review of our knowledge and some thought toward future inquiry. One principal thesis that emerges from this summary "view from midstream"—midstream on the Teays and midstream in our studies—is that we are too complacent in our approach to that part of geomorphology that hides beneath and around the edge of the drift cover. We too easily accept that the Teays is preglacial, that the Ohio is ice-marginal, and that there are too few drainage cycles and too few episodes of drainage rearrangement that accomplish too little. Thus one of my tasks here is to question the too-easy answers.

Gray, H. H., 1991, Origin and history of the Teays drainage system: The view from midstream, *in* Melhorn, W. N., and Kempton, J. P., eds., Geology and hydrogeology of the Teays-Mahomet Bedrock Valley System: Boulder, Colorado, Geological Society of America Special Paper 258.

WHAT IS THE AGE OF THE TEAYS VALLEY SYSTEM?

A major part of the buried Teays Valley system is shown on the bedrock topographic map of Indiana (Gray, 1982). Only a few years ago, most geologists would have accepted such a map as representing the preglacial topography of Indiana. We now know that this surface is complex, a mosaic showing only the bedrock bits and pieces of surfaces of many ages. What is next is to sort out the pieces, arrange them in order, and determine their ages. This will be an immense job, and many details may never be worked out.

From this midstream vantage point, however, we can look up and down the Teays Valley and ask some pertinent questions. First, is the Teays preglacial? This concept begins with the notion that most major geomorphic features outside what is conventionally recognized as the glaciated area are per se preglacial. Also involved is the thought that because the Teays is the oldest complete drainage system that we have been able to discern in this area, the entire system must be preglacial. These ideas are presumptive but simplistic; firmer grounds must be found.

The valley of the Teays across east-central Indiana (Fig. 1) is relatively straight and is 200 to 300 ft (60 to 90 m) deep. In places it is scarcely a mile (1.6 km) wide at the rim. Valley walls are steep and break abruptly from the adjoining plateau. There are few tributaries of much length; most are steep, and some appear to be hanging, that is, they appear not to join at grade. The walls of this deep valley are mostly limestone and dolomite, but karst development in the adjoining upland, as indicated by such features as water-producing, gravel-filled, solution-enlarged joints, is limited to the upper 50 to 100 ft (15 to 30 m) of the carbonate-rock sequence.

As described by Tight (1903) and most subsequent authors, the Teays has a drainage basin that compares in size with that of the present Ohio River, and although the valley of the Teays in eastern Indiana is as deep as that of the Ohio, it is narrower and

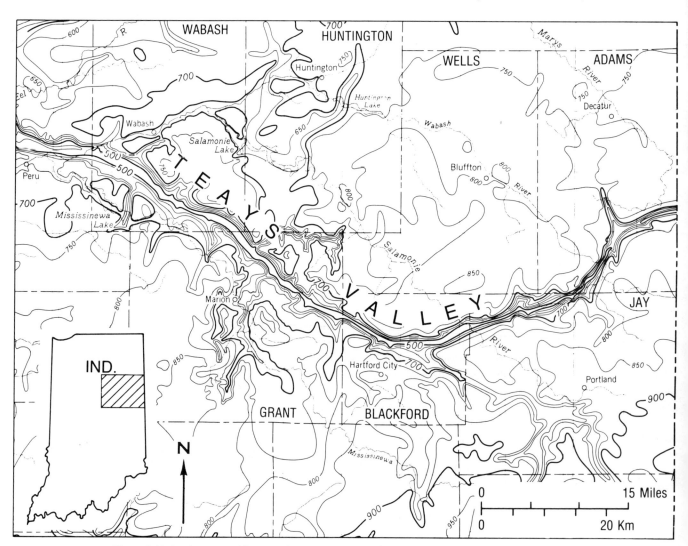

Figure 1. Bedrock topographic map of a part of east-central Indiana showing the gorge of the Teays Valley. Modified from Gray (1982). Contour interval is 50 ft (about 15 m).

has fewer tributaries and a less well-integrated drainage network. Where the Ohio crosses limestone terrane, it has the same kind of short, steep tributaries as does the Teays. None of them, however, are hanging with respect to the bedrock valley, and karst development adjacent to the Ohio is as deep as the bedrock valley, well below the present level of the river.

If the part of the Teays shown in Figure 1 is preglacial, if it is therefore a downstream extension of the very ancient New River of West Virginia, then it must have been many million years old at the time of its demise and the consequent creation of the Ohio River. This event took place about 0.8 Ma, according to present best estimates (see, for example, Fullerton, 1986, p. 24; other chapters, this volume). The Ohio, therefore, is relatively young, yet in many ways it appears more mature than the Teays. I therefore question that the buried part of the Teays system is preglacial, and instead suggest that the history of the Teays gorge across Indiana is shorter than that of the Ohio.

WAS THE TEAYS VALLEY SYSTEM CREATED BY GLACIAL DIVERSION?

If this segment of the Teays is not preglacial, then how might the course of the Teays have been affected by glaciation? Surface parts of the Teays system in southeastern Ohio and adjacent parts of West Virginia (volume frontispiece) exhibit a meandering and dentritic pattern that suggests a mature drainage system, but just northwest of Circleville in central Ohio the buried course of the Teays turns sharply westward, leaves the wide and deeply alluviated valley of the Scioto River, and enters a limestone-walled gorge. This gorge crosses a broad, high bedrock divide—one of the highest bedrock features in Ohio— and continues northwestward. Somewhere between Sidney and Celina, in west-central Ohio, a major northward-flowing tributary, the extended ancestral Kentucky River, joins. The main valley then swings westward, crosses a lesser divide, and enters Indiana. At this point it turns southwestward, crosses yet another divide through an extremely narrow and somewhat tortuous gorge (Fig. 1), and enters what appears to be a former integrated valley system that flowed northwestward down the regional slope of the bedrock.

Thus a segment of the Teays nearly 150 mi (250 km) long crosses the regional grain of the bedrock surface. I believe that the most probable explanation for this is that the Teays was diverted into this path by one or more of the early episodes of glaciation. I therefore suggest that the Teays Valley system be provisionally assigned at least three stages of development: a pre-diversion Teays I, which is perhaps truly preglacial and during which the upper Teays and the ancestral Kentucky were not deeply entrenched and may have flowed northward into the bedrock lowland that now underlies Lake Erie; Teays II, the integrated westward-flowing Teays system; and Teays III, the entrenched valley as we have come to know it. This entrenchment is so deep that it seems likely that the bedrock valley was graded to a glacially depressed sea level, as may also have been the later "Deep Stage" drainage that was recognized (but not so explained) by Stout and others (1943).

As an aside here, we need a better nomenclature to deal with the several "deep stages" that now are known, and clearer concepts of their causes and temporal relations. My present inclination is to associate the "deep stages" with the onset of major glacial events, associated climatic changes, and depression of sea level. There could be many such stages, but this concept, though attractive, remains to be rigorously tested. If, however, lowered sea level had no significant effect this far inland, then, because major advances of the different ice lobes now are known to be not necessarily synchronous, the "deep stages" may reflect severe periglacial climates in regions that were marginal to, but not directly affected by, active ice sheets.

PROBLEMS IN DEFINING BURIED VALLEYS

The part of the buried Teays Valley that is shown in Figure 1 has been known to oil and gas well drillers since the boisterous activity in the Trenton Field 100 years ago. Records were not kept for most of those wells, unfortunately, but newer data define this part of the valley quite well. That is not to say that its proper interpretation is not tricky; to define the thalweg of the valley, especially, is something like the old parlor game of pinning the tail on the donkey: the target is small and elusive.

Usually the thalweg must be drawn by threading it from one deepest control point some distance, through areas in which existing control does not adequately define the valley bottom but only shows where the thalweg is not, to a second definitive point. This practice involves negotiating a narrow line between two kinds of error that are similar to the two types of complementary error in statistics. Drawn too extravagantly, the thalweg will be shown in places where it does not belong; drawn too conservatively, it will give an inadequate concept of the depth and expectable continuity of the thalweg. There are a few places along the narrow stretch of the Teays in Indiana (Fig. 1) where drilling suggests that the thalweg is deeper than shown, but these indications are not adequately confirmed by data up-valley or down-valley.

Seismic data present their own problems in defining valley bottoms. Refraction seismic techniques tend to average the depths beneath the geophone string, so that high points are almost invariably interpreted too low and low points too high. The only way that refraction methods can properly identify the deepest part of a sharp valley is to position the geophone string precisely down the axis of the valley; in practice this is difficult to do. Furthermore, there is in some parts of northern Indiana a layer of till that is so hard and has such a high seismic response that it commonly is interpreted as bedrock. Gravity and resistivity data also are beset by similar difficulties.

Ultimately, however, the problem is a statistical one: how to define a thalweg that occupies far less than 1 percent of the landscape. Less than one control point in a hundred can be expected to be on target! Because of these difficulties, I do not attempt to profile the Teays Valley but point out that even (or especially!) with the best of data the valley floor appears to be somewhat bumpy and does not maintain a uniform gradient. This suggests that it is premature to look for knickpoints, to propose crustal movements, or to make hydrologic calculations based on perceived gradient, except in a very preliminary way.

THE ANCESTRAL KENTUCKY RIVER: A TRIBUTARY TO THE TEAYS?

A major tributary to the Teays system from the south is the ancestral Kentucky River. It has often been suggested that the ancestral Kentucky flowed somehow into southeastern Indiana, but this is improbable. Bedrock walls enclose the lower Whitewater valley (Fig. 2), and the bedrock floor of the valley rises steeply upstream. This implies that the modern Whitewater and its major branches originated with the Ohio and therefore are younger than the Teays and the ancestral Kentucky. A deeply buried and somewhat enigmatic valley system in Henry County and western Wayne County, here designated the New Castle Valley, apparently did flow northwestward to join the equally deep Anderson Valley, a major westward-flowing tributary to the Teays (Fig. 2), but thresholds at the head of the New Castle Valley are 100 ft (30 m) or more too high to admit the passage of an extension of the ancestral Kentucky; they may simply be cols formed by the intersection of Deep Stage Whitewater valleys with preexisting buried Teays-age valleys.

That the present Miami valley once contained a northward-flowing extension of the ancestral Kentucky River has been disputed, first for lack of confirming data, and second, I believe, because of misunderstanding the kind of data needed. Subsequent to formation of the Ohio River, the Miami valley was occupied by the Hamilton River, which belongs to the Deep Stage of Stout and others (1944), although these authors seem not to have realized this. The deeply entrenched Hamilton flowed southwestward to join the deeply entrenched Ohio, first at a point north of Cincinnati, and later, after the Ohio was glacially rerouted, at the Indiana-Ohio state line; it may therefore be the survivor of at least two "deep stages" (see, for example, Teller, 1973).

At the mouth of the Miami, the bedrock valley floor is about 300 ft (90 m) below the level of the ancestral Kentucky and about 100 ft (30 m) below present drainage. If the ancestral Kentucky extended northward along the present Miami, the level of its northward-descending valley should converge northward on the younger southward-descending (Hamilton) bedrock valley floor. Valley remnants that would confirm this apparently have not yet been noted in Ohio, perhaps because they are obscured by younger glacial deposits, perhaps for lack of detailed maps such as are available for that part of Kentucky through which the ancestral Kentucky River and its tributaries have been traced (for example, Swadley, 1972), or perhaps because the evidence has been entirely destroyed by Deep Stage and later erosion. Deep drilling at dam sites north of Dayton, however, suggests a northward-descending extension of the ancestral Kentucky at least to that point (Teller and Goldthwait, this volume). Thus, this important tributary flowed northward across western Ohio to join the Teays in west-central Ohio (see frontispiece), as indicated by Swadley (1971), Teller (1973), and others, although much of its route, especially the northern part, is not yet well documented.

WAS THE OHIO RIVER CREATED AS AN ICE-MARGINAL STREAM?

From Lawrenceburg, Indiana, to Carrollton, Kentucky, the southwestward-flowing Ohio River closely parallels, but does not occupy, the northeastward-draining meandering valley of the ancestral Kentucky (Fig. 3). The Ohio's course has commonly been considered to be ice-marginal, a supposition that seems fortified by map interpretations showing Illinoian and older drift boundaries essentially at the river. (See, for example, Campbell and others, 1974.) Because of differing philosophies of mapping, however, older drifts are more widespread in Kentucky and less widespread in Ohio than most published maps would lead one to expect. Boundaries of the older drifts are not at the river, and other possible hypotheses for the origin of the Ohio should be considered.

Since the time of Tight, the conventional concept has been that much of the drainage rearrangement that ultimately resulted in creation of the Ohio River took place on an extensive lacustrine or alluvial surface that was created following glacial blockage of Teays III. Such a surface has been conjectured for southeastern Ohio; a similar present surface is that of the Wabash Lowland in southwestern Indiana and adjacent parts of Illinois, with its vast alluvial-lacustrine flats and loess-covered, bedrock-cored island hills.

The old Kentucky valley indeed contains remanent slackwater deposits, which rise to an altitude of about 730 ft (220 m; Swadley, 1971). This level is well below the surrounding upland and is concordant with deposits in the Licking-Claryville Valley of Luft (1980) and of Teller and Last (1981), which is just to the east and which was tributary to the extended ancestral Kentucky. These midlevel deposits are very limited in extent today, owing to extensive dissection, but even at a maximum postulated former extent it seems unlikely that they were sufficiently widespread to have allowed an Ohio, swollen with glacial meltwater as it was, free rein in establishing a new and nearly independent course.

Indeed, much of the area that now is below the critical 730-ft (220 m) level has obviously been cut not by the ancestral Kentucky but by the Ohio itself (Fig. 3). And so, taking into consideration both the rectangular pattern of the Ohio's course in this reach and its independence of the course of the older stream, it seems plausible that here the Ohio may have been superposed, either from jointed ice or from a thin till cover, or it may in part have originated as a subglacial stream flowing under hydraulic head. An analogous situation upstream from Cincinnati is differently explained, however, by Teller and Last (1981).

Whatever its mechanism, the date of this event remains obscure. Not only are the slackwater sediments of the old Kentucky system about 100 ft (30 m) lower than those in most other parts of the disrupted Teays (see other chapters, this volume),

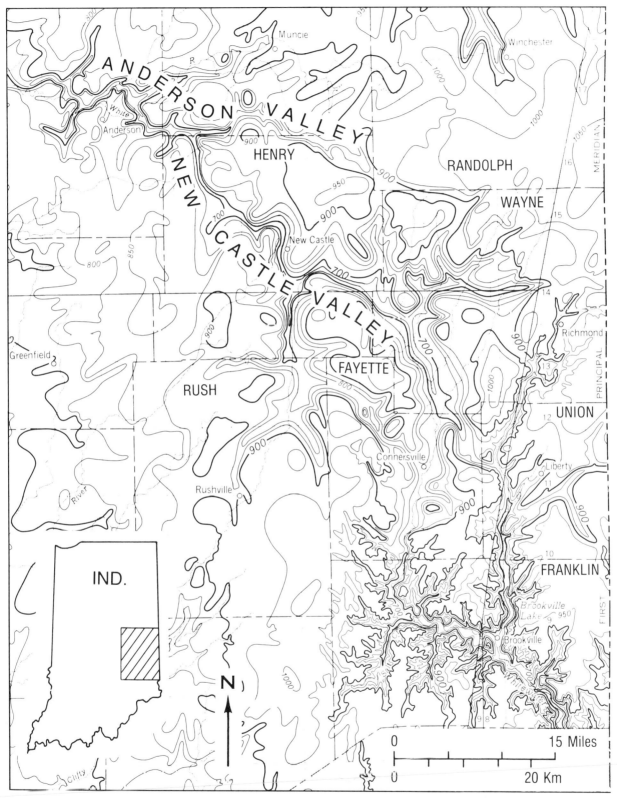

Figure 2. Bedrock topographic map of a part of southeastern Indiana showing valleys of the modern Whitewater River and the buried New Castle and Anderson Valleys. Modified from Gray (1982). Contour interval is 50 ft (about 15 m).

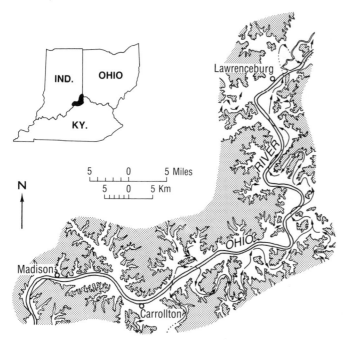

Figure 3. Map showing a part of the course of the ancestral Kentucky River and a few of its tributaries (arrows; mostly after Swadley, 1971) in relation to the course of the Ohio River from Lawrenceburg to Madison, Indiana. Shading indicates areas that are above the critical 730-ft (220 m) level to which the ancestral system was filled by lacustrine deposits sometime prior to creation of this segment of the Ohio River.

RELATION OF THE BLUE RIVER STRATH

In south-central Indiana, on the Blue River and on other streams that drain primarily limestone terrane, there is a prominent karsted strath surface that is about 100 ft (30 m) above present drainage and 300 ft (90 m) below the roughly accordant nearby summits. Remnants of this surface include dry tributary valleys and abandoned valley segments that owe their preservation to extensive development of underground drainage subsequent to the formation and entrenchment of the Ohio River. This surface was named the Blue River Strath by Powell (1964), who correlated it with the Parker Strath and who pointed out that the Blue River Strath merges upstream with the karst surface of the Mitchell Plain, itself also a relict.

In places on the Mitchell Plain and on the Blue River Strath are patches of gravel consisting of water-worn pebbles of chert, geodes, and sandstone. These appear to be equivalent to the Luce Gravel of the Owensboro, Kentucky, area, which Ray (1965) named and described as "bronzed" because of the dark coatings of iron oxide on the pebbles, and to the similar Mounds Gravel (Willman and Frye, 1970) of Illinois. Thus, both position within the landscape and nature of associated deposits support a speculative time correlation of the Blue River Strath with the ancestral Kentucky and the type Teays. The Blue River Strath has not been widely recognized, possibly because of its apparent dependence on limestone terrane, but a similar strath surface is present in Kentucky on the Salt River.

SOME ADDITIONAL PROBLEMS

To return to the Teays Valley: in northern Indiana it has two distinct parts—an eastern gorge-like segment already discussed (Fig. 1), and a more complex, branching, broader valley to the west (Fig. 4). The western segment is multicyclic; its history of erosion and filling differs greatly from that of the gorge and is instead similar to the history of the Mahomet Valley of Illinois (Bleuer, 1989). I think this illustrates that we too easily accept that the Teays is the Teays is the Teays; it is not, and later reoccupations of this western valley system should be given their own distinct names. Indeed, we should now take another next step: differentiate and specifically name segments of the valley itself so as to make clear when we are talking about the history of the Teays and its drainage network and when we are talking about the evolution of a specific valley segment. These arguments are further discussed by Bleuer in this volume.

A final hypothesis: if there are within the glaciated area vestiges of a truly preglacial drainage network, that is, remnants of a pre–Teays II system, they should be looked for north of the reoriented Teays where they may have been preserved beneath deposits of the early glacier (or glaciers) that I believe diverted Teays I into a westward course. A possible candidate is a major valley system that heads in northeastern Indiana (frontispiece), extends across northwestern Ohio and southeastern Michigan, and enters the bedrock lowland beneath Lake Erie along an axis

they are also magnetically normal, not reversed (Teller and Last, 1981). This implies that they are younger, and for this reason, Fullerton (1986, p. 27), without discussing possible effects on drainage patterns, related the Kentucky valley slackwater fill to a glacial event some 0.2 m.y. later than that which closed major parts of the Teays elsewhere. I find it difficult to understand how this part of the course of the Ohio could be significantly younger than upstream parts that seem securely dated at 0.8 Ma, but I have no good alternative to suggest.

At this point it is well to recall how much other fragmental evidence indicates how little we know about the early glaciations in the Midwest. Erratic boulders are scattered across supposedly unglaciated parts of northern Kentucky (Leverett, 1929), and well beyond the mapped glacial boundary in eastern Ohio are isolated high-level sediments that contain erratic material (Lessig, 1963). Sharp elbows on the Salt River near Lawrenceburg, Kentucky, and similar anomalies along the type Teays itself more readily signify diversion than capture. These selected occurrences, all of which are far beyond conventionally mapped glacial boundaries, suggest that there are older glaciations in this region that we know almost nothing about. What does the Ohio owe to them? At this point the question must be asked, but we are not yet prepared to answer.

that differs considerably from the axis of the present lake and its immediate predecessors. This valley is as yet poorly defined, however, and we know nothing of its fill; I therefore can only suggest that it may predate, and probably is no younger than, the integrated but not yet entrenched Teays II. The narrower valley that drained from northeastern Indiana southwestward into the Teays, the Metea Valley of Wayne (1956), is much younger and is associated with the Mahomet, rather than with the Teays system.

Quaternary geology is a field in which the corpus of knowledge is growing rapidly and concepts are changing with equal speed. In particular, subsurface data and techniques applicable to subsurface studies have become nearly mind-boggling in diversity and scope. Yet there is a need now to pause and to assemble what is known from one end of the Teays system to the other. This symposium and this volume are a first attempt to fill that need, but much more remains to be done before we can perceive the entire elephant rather than merely its parts.

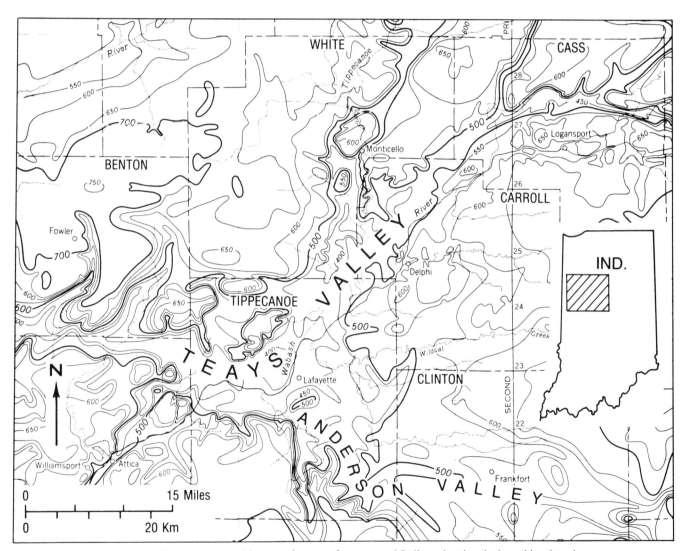

Figure 4. Bedrock topographic map of a part of west-central Indiana showing the branching, broad Teays (Mahomet) Valley and a major tributary, the Anderson Valley. Modified from Gray (1982). Contour interval is 50 ft (about 15 m).

REFERENCES CITED

Bleuer, N. K., 1989, Historical and geomorphic concepts of the Lafayette Bedrock Valley System (so-called Teays Valley) in Indiana: Indiana Geological Survey Special Report 46, 11 p.

Campbell, L. J., and others, 1974, Late Cenozoic geologic features of the middle Ohio River valley: Geological Society of Kentucky Guidebook, 25 p.

Cummins, J. W., 1959, Buried river valleys in Ohio: Ohio Division of Water, Water Plan Inventory Report 10, maps.

Fidlar, M. M., 1948, Physiography of the lower Wabash Valley: Indiana Division of Geology Bulletin 2, 112 p.

Fullerton, D. S., 1986, Stratigraphy and correlation of glacial deposits from Indiana to New York and New Jersey, in Sibrava, V., Bowen, D. Q., and Richmond, G. M., eds., Quaternary glaciations in the Northern Hemisphere: Oxford, England, Pergamon Press, p. 23–37.

Grant, R. P., and Pringle, G. H., 1943, Topographic map of the bed rock surface of the southern peninsula of Michigan: Michigan Geological Survey.

Gray, H. H., 1982, Map of Indiana showing topography of the bedrock surface: Indiana Geological Survey Miscellaneous Map 35, scale 1:500,000.

Horberg, L., 1945, A major buried valley in east-central Illinois and its regional relationships: Journal of Geology, v. 53, no. 5, p. 349–359; Illinois State Geological Survey Report of Investigations 106, 11 p.

—— , 1950, Bedrock topography of Illinois: Illinois Geological Survey Bulletin 73, 111 p.

Lessig, H. D., 1963, Calcutta Silt, a very early Pleistocene deposit, upper Ohio Valley: Geological Society of America Bulletin, v. 74, p. 129–140.

Leverett, F., 1929, The Pleistocene of northern Kentucky: Kentucky Geological Survey, Series VI, v. 31, p. 1–80.

Luft, S. J., 1980, Map showing the late preglacial (Teays-age) course and pre-Illinoian deposits of the Licking River in north-central Kentucky: U.S. Geological Survey Miscellaneous Field Studies Map MF-1194, scale 1:125,000.

McGrain, P., 1950, Thickness of glacial drift in north-central Indiana: Indiana Flood Control and Water Resources Commission Circular 1, map.

Powell, R. L., 1964, Origin of the Mitchell Plain in south-central Indiana: Proceedings Indiana Academy of Science, v. 72, p. 177–182.

Ray, L. L., 1965, Geomorphology and Quaternary geology of the Owensboro Quadrangle, Indiana and Kentucky: U.S. Geological Survey Professional Paper 488, 72 p.

Stout, W. E., Ver Steeg, K., and Lamb, G. F., 1943, Geology of water in Ohio: Ohio Geological Survey, 4th series, Bulletin 44, 694 p.

Swadley, W C, 1971, The preglacial Kentucky River of northern Kentucky: U.S. Geological Survey Professional Paper 750-D, p. D127–D131.

—— , 1972, Geologic map of parts of the Lawrenceburg, Aurora, and Hooven Quadrangles, Boone County, Kentucky: U.S. Geological Survey Map GQ-989, scale 1:24,000.

Teller, J. T., 1973, Preglacial (Teays) and early glacial drainage in the Cincinnati area, Ohio, Kentucky, and Indiana: Geological Society of America Bulletin, v. 84, p. 3677–3688.

Teller, J. T., and Last, W. M., 1981, The Claryville Clay and early glacial drainage in the Cincinnati, Ohio, region: Palaeogeography, Palaeoclimatology, Palaeoecology, v. 33, p. 347–367.

Tight, W. G., 1903, Drainage modifications in southern Ohio and adjacent parts of West Virginia and Kentucky: U.S. Geological Survey Professional Paper 13, 111 p.

Wayne, W. J., 1956, Thickness of drift and bedrock physiography of Indiana north of the Wisconsin glacial boundary: Indiana Geological Survey Report of Progress 7, 70 p.

Willman, H. B., and Frye, J. C., 1970, Pleistocene stratigraphy of Illinois: Illinois Geological Survey Bulletin 94, 204 p.

MANUSCRIPT ACCEPTED BY THE SOCIETY JUNE 29, 1990

The Lafayette Bedrock Valley System of Indiana; Concept, form, and fill stratigraphy

N. K. Bleuer
Geological Survey Division, Indiana Department of Natural Resources, 611 North Walnut Grove Avenue, Bloomington, Indiana 47405

ABSTRACT

The Lafayette Bedrock Valley System is a complex of bedrock valleys that converge on and diverge from Lafayette, Indiana. The primary trunk valley of the system, composed of the narrow Marion Valley Section on the east and the broad Mahomet Valley Section on the west, is the classic "Teays Valley" of the Midwest. If such a continuous Teays drainage truly existed, it represents only the early part of the history of the Lafayette Bedrock Valley System in Indiana.

Origins of the valley parts and their form remain enigmatic. Although the Marion crosses major rock structure, the course and the contrasting forms of the parts reflect structural and lithologic control. Contrasting valley forms, valley deeps, and possible inset benches may reflect one event or multiple events in a single valley, or disparate events in other valleys; or the features may reflect external events, such as incision through forebulge or erosion beneath bursting ice dams.

Although the origins of the valley system are conjectural, the fill sequences within give evidence of the nature and timing of the demise of the system. The Marion valley is filled with a plug of old lacustrine and glaciolacustrine sediments included in the Blackford Member of the Banner *and* Jessup Formations. These sediments were deposited in a lake dammed between ice at the valley bends at Logansport, Indiana, and St. Marys, Ohio. Deposited at the dams were subaqueous fan deposits of coarse-grained outwash and tills of both basal-meltout and sediment gravity-flow origin. The tills include the red claystone-bearing West Lebanon Till Member on the west, and the Wilshire Till Member on the east. An uppermost tongue of the West Lebanon till caps the Blackford lacustrine sediments, indicating southeastward progression of the West Lebanon ice into the lake and ultimately over the entire fill sequence. The relative age of plugging of this valley section is suggested by the West Lebanon, which overlies magnetically reversed ($>$0.7-m.y.-old; marine isotopic stage 22?) sediments in western Indiana. The plugging of the Marion valley by West Lebanon ice corresponds in time with the plugging of the valley in Ohio by Wilshire ice (and the deposition of the Minford Silts) and marks the end of classic Teays-stage regional drainage.

The Mahomet valley subsequently was reexcavated as part of a Metea-Mahomet drainage system, heading in northeastern Indiana. This valley is filled with younger, interfingered outwash and till, which are included in the Mahomet Member and the Brookston Till Member of the Banner and Jessup Formations. These deposits represent aggrading braided stream and fan environments in front of southwestward-advancing Brookston ice. The relative age of plugging of this valley section is given by the Vandalia Till Member of the Glasford Formation, an Illinoian till that caps the valley fill, and by

Bleuer, N. K., 1991, The Lafayette Bedrock Valley System of Indiana; Concept, form, and fill stratigraphy, *in* Melhorn, W. N., and Kempton, J. P., eds., Geology and hydrogeology of the Teays-Mahomet Bedrock Valley System: Boulder, Colorado, Geological Society of America Special Paper 258.

the West Lebanon till, which was apparently cut out prior to valley filling. The final plugging of the Mahomet marks the end of any deeply incised drainage in north-central Indiana.

With the demise of the Mahomet drainage outlet, development of an upper Wabash drainage system began. The fill of bedrock valleys south of the Marion-Mahomet trunk valley contains evidence of multiple erosional surfaces. The gradients of these surfaces suggest a merging at Lafayette into early equivalents of the modern Wabash drainage, exiting into the Wabash bedrock valley via the Attica cutoff or across the rock sill above Independence.

INTRODUCTION

The fill of the so-called Teays Valley across Indiana remains as an isolated time capsule containing the stratigraphic record of interactions of the earliest northern and eastern ice lobes (Bleuer, 1980). That fill contains a record of cutouts and insets associated with drainage development in cross-cutting valleys. Not surprisingly, the record here parallels classic ideas of the history of the Teays Valley, which have derived partly from the crosscutting and inset relations of younger drainage with respect to it (frontispiece; Tight, 1903; Stout and others, 1943). The type Teays and the deposits within it hang above the modern valleys of the Kanawha and Ohio Rivers in West Virginia and above the valley of the Miami River in Ohio (cover photo, this volume).

Fill stratigraphy records nothing of the early development of the trunk valley, the classic Teays of Horberg (1945). However, it does record the demise of any through-flowing drainage system, and development of subsequent Midwestern drainage. Therefore, I include here only an introductory discussion of concepts of the rock valleys and drainage systems, and a description of the form of the trunk valley. The early development of these valleys and concepts of the term "Teays" are discussed more fully elsewhere in this volume (see Goldthwait; Gray; Kempton and others; and Melhorn and Kempton). I concentrate, instead, on the fill stratigraphy of the trunk valley and related valleys of the Lafayette system in Indiana.

This chapter derives from the Indiana Department of Natural Resources (DNR) Teays project, a joint project of the Division of Water and the Division of Geological Survey (IGS), for which a contract rotary drilling program was completed during 1978 and 1979. This was the first large-scale effort toward sampling and downhole gamma-ray logging of thick glacial sequences in Indiana. Since that time, hundreds of additional sample-log sets have been obtained across glaciated Indiana, and much has been learned about the log characteristics of glacial vertical sequences. Log interpretation is treated only casually here; preliminary discussions of this work are in Bleuer (1986), Bleuer and Fraser (1988, 1989), and Bleuer and Melhorn (1989).

CONCEPT OF THE LAFAYETTE BEDROCK VALLEY SYSTEM

The terminology applied to the deep rock valley that extends east-west across north-central Indiana has evolved in parallel with concepts of its route (Table 1, Fig. 1). Because

TABLE 1. DEVELOPMENT OF TERMINOLOGY FOR THE LAFAYETTE BEDROCK VALLEY SYSTEM IN INDIANA*

Terminology	Outlet Route	Reference
Ancient valley of the Wabash	Lower Wabash valley	Phinney, 1890
Deep drive, Loblolly	Southward	Blatchley, 1897
Preglacial Wabash	Lower Wabash valley	Leverett, 1895; Dryer, 1920
Deep preglacial channel	Little Miami River Valley	Bownocker, 1899
Teays Valley	Lower Wabash valley	Fidlar, 1943
Lower Teays	Mahomet valley	Horberg, 1945
Kanawha (Teays)	Mahomet valley	Thornbury, 1948
Teays (Kanawha)	Mahomet valley	McGrain, 1950
Mahomet-Teays	Mahomet valley; lower Wabash valley via Danville valley	Wayne, 1952, 1956
Lafayette Bedrock Valley System	Mahomet valley; lower Wabash valley via Danville valley and Attica cutoff	Bleuer, 1989; this chapter

*Summarized from Bleuer, 1989.

connections and routes have been deduced from relatively little data, concepts have depended on interpretive necessities of the time, such as draining any valley out of Indiana via the only route apparent, either an ancestral Wabash River valley or the ancestral Little Miami River valley; draining a Teays Valley in Ohio into Indiana via the only route apparent, the Loblolly; and draining *something* into the Mahomet Bedrock Valley of Illinois via the only route apparent, the Teays. The present concept of the Teays in the Midwest began in this manner with Horberg's (1945) connection of the Mahomet valley of Illinois with the Teays of Indiana, as mapped by Fidlar (1943). Finally, a variety of drainage routes were hypothesized, which developed sequencially within those valleys in response to glaciation (Wayne, 1952; Bleuer and others, 1982, 1983; Bleuer, 1980, 1989).

In fact, the deep bedrock valley extending east-west across north-central Indiana is but the trunk valley of a larger system of rock valleys that converge on and diverge from the Lafayette area (Wayne, 1952, 1956; Gray, 1982, and this volume; Bruns and others, 1985; Bleuer, 1989). Some parts are shallow notches superposed across rock uplands; these are associated with the modern Maumee-Wabash Trough (Fig. 1) and represent events distinctly younger than events that shaped the deep parts of the system. But most of the rock valleys and lowlands are deep, with floors merging with that of the trunk valley itself.

With many possible routes for entry and exit of drainage about the Lafayette hub, it is presumptuous to consider, a priori, that the bedrock valley system existed in present form at any single time in the past. Therefore, a distinction needs to be made between elements of the physically defined bedrock *valley* system and the conceptual *drainage* systems that are interpreted to have flowed within those elements. For this reason, the system of bedrock valleys of north-central Indiana has been termed the "Lafayette Bedrock Valley System" (Bleuer, 1989, in an elaboration of names and concepts of Wayne, 1956).

The apparent trunk valley of the Lafayette system, that which heretofore has borne the name "Teays Valley" in Indiana, is composed of the Marion Valley, Logansport Bend, Battle

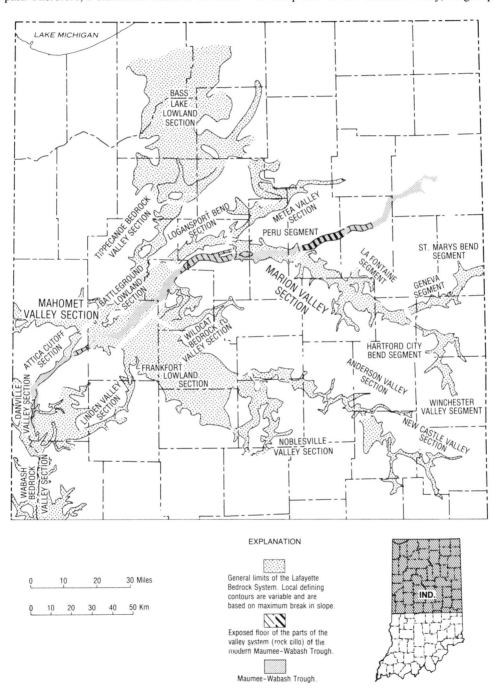

Figure 1. Terminology applied to the Lafayette Bedrock Valley System and the Maumee-Wabash Trough, north-central Indiana (modified from Bleuer, 1989).

Ground Lowland, and Mahomet Valley Sections (Fig. 1, Table 1); it is commonly referred to below as simply the "trunk valley" of the Lafayette system. Discrete parts of the Marion valley include the Peru, La Fontaine, Geneva, and St. Marys Bend Segments. Primary tributaries joining the trunk at or above Lafayette include the Metea Valley, the Tippecanoe Bedrock Valley, and the Frankfort Lowland Sections, and tributaries to the lowland, the Anderson and New Castle Valley Sections. Exit valleys west and southwest of Lafayette include the Mahomet Valley Section and connections to the Wabash valley that include the Danville Valley, Attica Cutoff, and Linden Valley Sections, and the shallow Independence sill of the modern Maumee-Wabash Trough (following the formal terminology of sections and segments of Bleuer, 1989, Plate 1 and Table 1).

FORM AND ORIGINS OF THE TRUNK VALLEY

Early concepts

Early descriptions of parts of the Marion-Mahomet trunk valley (Table 1; Bleuer, 1989) were colorful and generalized, such as "great basin" (Gorby, 1886), the "deep drive" of the Loblolly district (Blatchley, 1897), or "deep preglacial channel" (Bownocker, 1899). The earliest generalized bedrock contours suggested the deep, narrow form of the Marion valley (Capps, 1910, Plate 2), in contrast to the broader central and western valley (Horberg, 1945, Fig. 4).

Valley form became more apparent through drift-thickness mapping (Wayne and Thornbury, 1951; Thornbury and Deane, 1955), which culminated in McGrain's (1950) cross-state swath focused on the Teays, and finally in Wayne's (1956) massive map compilation of drift thickness in central and northern Indiana. In Indiana, the Marion was characterized thenceforth as a "nearly vertical walled chasm . . . that developed because the massive Niagaran dolomites and limestones in the walls of the valley were more resistant to erosion than were the Cincinnatian shales . . . at the bottom of the valley. . . . Niagaran reefs probably stood like bastions in the valley walls just as they do today where they have been exhumed by present rivers," and "the resistance of reef rock to erosion may have caused unusual valley constriction" (Wayne, 1956, p. 37). This description corresponds with the "gorge-like" cross-sectional profile suggested by drilling traverses in west-central Ohio (Norris and Spicer, 1958, p. 218); in fact, in a boring near St. Marys, Ohio, alternating limestone ledges and drift between depths of 182 and 330 ft have been interpreted as a vertical valley wall (Bownocker, 1899).

While early work on the Mahomet valley in Illinois concerned stratigraphy and form (Horberg, 1945), work in Indiana stressed geomorphologic interpretations (Wayne and Thornbury, 1951; Wayne, 1956), obviously owing to the challenge of earlier Ohio literature. The latter dwelt on the relations of the type (exposed) Teays Valley, the Parker Strath, and valleys incised into the Teays valley floor that were termed "deep-stage." Valley profiles (Table 2) were cited, I suspect, as evidence of a continuity of the buried sections with the exposed valley of the upper Teays, which, in contrast to descriptions of the buried sections, had been considered "quite mature as shown by the low gradient, the broad valley in which it flowed, and by the bordering hills well rounded and degraded" (Stout and others, 1943). Wayne and Thornbury (1951) recognized an intermediate, or strath, surface in Wabash County, but Wayne (1956) concluded that the entire depth of what had been termed "Teays" in Indiana was, indeed, the Teays, not deep-stage.

Overall valley form is now best shown on large-scale, regional bedrock topographic maps that appeared following the completion of U.S. Geological Survey 7½-min topographic mapping in Indiana (Burger and others, 1966; Gray, 1982). The most detailed mapping of the trunk valley (Bruns and others, 1985) followed a concentrated program of field checking of water wells and acquisition of geophysical data, which constituted the initial phase of the DNR project. This and Gray's (1982) map allow for the first legitimate comparison of the parts of the Lafayette system.

Interpretation of modern data

Prior to the DNR study, the thalweg of the trunk valley could be located only with an accuracy of plus or minus several miles almost anywhere in the state. Hundreds of records are either unlocatable (old Trenton Field), or record minimum depths to rock, or show where a valley is not. Because specific test-drilling sites could not be located, new datum points were generated from end-line reverse seismic refraction studies by Joseph Whaley, IGS, along over 50 lines of cross section. (Lines are plotted in Bruns and others, 1985.)

These seismic lines provide the only real cross sections of valley form, but they are imperfect. (See the related discussion in Gray, this volume.) In most lines across the Marion valley, the

TABLE 2. GRADIENTS OF THE MARION-MAHOMET TRUNK VALLEY OVER VARIOUS STRETCHES CALCULATED BY VARIOUS AUTHORS

Gradient (in./mi)	Location (Author)
12	Across Ohio (Stout and others, 1943)
8	Across Ohio and Indiana and into the Wabash Valley (Fidlar, 1943)
7	Across Ohio and Indiana to Beardstown, Illinois (Horberg, 1945)
2	Across Ohio to the supposed strath surface of Wabash County, Indiana (Wayne, 1956, referring to the unlikely correlation of Thornbury and Wayne, 1951)
12	Across Ohio to Wabash County, Indiana, including anomalously low elevations in Wabash County (Wayne, 1956)
9–10	Across Ohio (Norris and Spicer, 1958)
4.5–7	This chapter (see Fig. 1) or 0-12-0 (separate segments)

form of the valley floor remains unknown. A flat floor (two or more similar elevations) was not detected, although geophone spacings of 50 and 150 ft along a 1,350-ft line could have detected a flat valley floor as narrow as about one-quarter mile. Throughout much of the broad Mahomet valley, refraction data values are impossibly deep. Thick, dry gravel (dry, owing to proximity to the Wabash River Valley) limits energy input, and shale bedrock causes hidden-layer problems. Narrow, incised single or multiple channels were not and cannot be detected by seismic refraction. For this reason, about 20 to 50 ft of depth may have been consistently missed in seismic lines and in actual drilling within the Marion valley near the Ohio state line and immediately southwest of Wabash, and within the westernmost Mahomet valley, as will be discussed.

In both eastern and western sections of the valley, the gravity method locally appears to be of greater value than seismic refraction in delineating drift-filled valleys (King, 1974; Adams and others, 1975; Kayes, 1979; Fig. 3A, this chapter), but even this method is incapable of detecting narrow, inset channels. In addition, shallow seismic reflection methods may aid in delineating not only bedrock form, but major units within the valley fill as well (Mitchell, 1984).

Characteristics of the sections

Marion Valley Section. The easternmost part of the valley, the Marion Valley Section, cuts through carbonate strata atop the Cincinnati Arch, and in the eastern reach, into Ordovician shales. The courses of the linear elements of the Marion and Anderson sections clearly bisect regional joint trends in the bedrock (Fig. 2). The surrounding upland landscape is relatively flat.

Valley-defining contours descend east-to-west. Through the eastern carbonate tableland, the valley is sharply defined by the 750-ft contour (Fig. 3D). To the west, 700-ft contours spread away from a sub-500-ft inner trench. The trench is commonly 1 mi to 0.5 mi wide at the rim, and apparently is less than 0.25 mi wide at the floor (Fig. 3B; Gray, 1982; Bruns and others, 1985). Locally, the thalweg either is undetectably narrow or simply is not flat. The actual rim-to-floor relief of the rock valley is about 250 ft, and nowhere more than about 300 ft, as defined by the rim-defining 750-ft contour and the base-defining 500-ft contour (Fig. 3D). (The absolute definition of valley, and thus the definition of its overall form, depends on the map and the defining contours chosen. Several different sources are used for figures here: Gray, 1982, for Fig. 1; Indiana Geological Survey, 1983, for Fig. 5; Bruns and others, 1985, for Figs. 3D and 15.)

No obvious intermediate surfaces exist within the eastern valley according to recent work (Gray, 1982; Bruns and others, 1985; Fig. 3B), except for the sub-700-ft broadening of the upland landscape (Fig. 3D). A 600-ft intermediate surface in Wabash County (Wayne and Thornbury, 1951) does not appear in the original work (Wayne and Thornbury, 1951, Plate 6), except as a product of the contouring process.

Earlier literary characterizations of the form of the Marion valley probably are true in part, but overstated in general. The great thickness of the drift within and atop the valley in Indiana (>400 ft locally), the visual appearance of the stacked contours of the valley (Gray, 1982), and the extreme vertical exaggeration used to illustrate it (Fig. 3) suggest that the Marion is a monumental gorge; it is not. The floor of that "gorge" in Ohio is 3,000 ft wide, and side slopes are only about 1:9 (from data of Norris and Spicer, 1958). Side slopes in Indiana could be interpreted as shallower than those in Ohio; the valley's rim-to-floor depth is not unlike parts of the present Ohio or Wabash valleys.

Logansport Bend, Battle Ground Lowland, and Mahomet Valley Sections. The forms of central and western parts of the valley, the Logansport Bend, Battle Ground Lowland, and Mahomet Valley Sections, differ greatly from the Marion section. The valleys are cut through shale lowlands and across the margin of a shale-sandstone upland (Fig. 2).

The Logansport bend and upper Battle Ground lowland apparently contain multiple thalwegs, and rock rims are not eas-

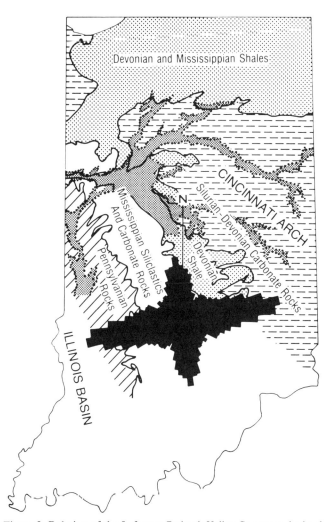

Figure 2. Relation of the Lafayette Bedrock Valley System to bedrock lithology and regional joint systems (bedrock adapted from Indiana Geological Survey, 1979; joint system after Ault, 1989, "combined measurements" from "area A" of his map).

ily defined. The overall landscape appears of more variable relief and dissection within a broadly sloping 700- to 550-ft flanking surface (Fig. 3D).

The form of the Mahomet resembles that described in Illinois (Kempton and others, this volume), a broad lowland surrounded by dissected, rolling uplands (Gray, 1982; Bruns and others, 1985; Fig. 3A). The 550-ft contour defines a diffuse valley rim. The valley appears to contain both 400- and 500-ft surfaces similar to benches in Illinois, and broad, sub-400-ft deeps appear to exist at and west of Lafayette (Fig. 3A; Kayes, 1979, from gravity profiles run after DNR drilling).

The form of the southern outlets of the western trunk valley is significant in the following discussion of valley-fill stratigraphy. The form of the Maumee-Wabash Trough across the modern Independence sill is obvious at the surface (Bleuer, 1989; USGS Attica and West Point 7½-min topographic quadrangles). The sill consists of rock outcrops across the valley in the Fulton Islands area, about 3 mi above Independence. The elevation of the channel there lies at about 490 ft. The exit elevation for the Attica cutoff, as given by water wells and bridge boring records at the buried valley's intersection with the Wabash at Attica, is about 400 ft, although the 450-ft contour is the lowest that has been

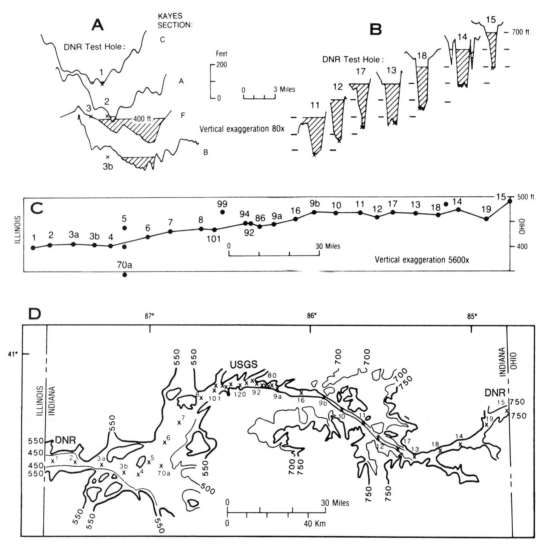

Figure 3. A, Cross sections of the Mahomet Valley Section, western Indiana, interpreted from gravity profiles A, B, C, and F of Kayes (1979, Figs. 62-64, 67). Projected rock-surface elevations of numbered DNR test holes are indicated by X. Elevations below 400 ft are shaded. B, Cross sections of the Marion Valley Section, eastern Indiana, interpreted from IGS seismic-refraction profiles. Rock-surface elevations determined from DNR test holes on line of sections are numbered and indicated by X. Elevations under 700 ft are shaded. C, Longitudinal profile of the base of the Marion-Mahomet trunk valley. Data are from DNR test holes and other selected drilling data. (Refer to Fig. 4 for test-hole locations.) D, Primary defining contours (in feet) for the Marion-Mahomet trunk valley, including locations of DNR and USGS test borings. (Refer to Fig. 6, below, for complete order of the latter; base map is adapted from Bruns and others, 1985).

continued through the valley thus far (Bruns and others, 1985). These elevations define the minimum depths at which sediments within the trunk upstream might relate to drainage exiting these routes.

Longitudinal profile of the trunk valley

Prior to the DNR study, concepts of the valley's profile depended on several old reports of floor depths that have become repeatedly cited classics. These reports, and various mapped contours based on them, are here ignored, as they probably are erroneous.*

The 500-ft and 400-ft contours are the highest and lowest presently mapped in eastern and western Indiana, respectively (Gray, 1982; Bruns and others, 1985). However, the elevation of the valley floor in easternmost Indiana is assumed to be as low as about 460 ft, given by data at St. Marys, Ohio (Stout and others, 1943). Most of the expanse of the valley floor exiting Indiana lies at about 400 ft, given by our test holes and corroborated by a gravity survey crossing the site of the DNR test hole 1 at Ambia (Fig. 3A; Kayes, 1979).

A broad deep lies below the 400-ft level at and west of Lafayette. Rock-floor elevations of about 360 and 376 ft are given by tests for the city of West Lafayette (test hole No. 70a) and for Purdue University (test hole No. 85-3). Earlier data (Rosenshein, 1958) includes numerous floor elevations just below 350 ft north and south of Lafayette. Gravity sections corroborate the presence of broad deeps nearby (Fig. 3A), but give no evidence of an exit into either the Mahomet valley or into the Wabash bedrock valley via the Attica cutoff.

From this information, the overall gradient of the Marion-Mahomet across Indiana must be between 4.5 and 7 in/mi (Table 2). Although this gradient appears to vary (Fig. 3C), this cannot be proven.

Origins of valley form

Whatever forces are called on to route the trunk valley across, rather than around, the Cincinnati Arch (see Gray, and Kempton and Melhorn, this volume), the ultimate routes of linear segments of the Marion valley (and parallel segments of the Anderson–New Castle, as well) appear to have been controlled by rock structure. The trends of the segments, bisecting trends of regional, primary joints, suggest drainage development along zigzagging, composite joint-controlled courses (Fig. 2). Such a composite (joint-trend bisecting) course characterizes most modern streams draining the west flank of the arch in southeastern Indiana.

The trunk valley transects three disparate landscapes (Fig. 3D) that suggest the nature of pre-"Teays" topography and drainage. An eastern trench cuts a flat >750-ft upland surface. (That surface has been glacially planed many times since the plugging of the valley and may not reflect preglacial topography at all.) A central continuation of that trench cuts a flat, <700-ft flanking bench. The western valley is itself a broad, <500-ft bench. These landscapes are clearly related to bedrock locally (Fig. 2), and the lower benches are, in fact, continuations of the broad, regional shale lowlands of northeastern Indiana. These lower surfaces must represent elements of early drainage systems that were finally welded into the so-called Teays. (See Gray, 1982, and Gray's discussion of pre-Teays II drainage, this volume.)

Similarly, valley-floor elevations may represent profiles of multiple surfaces that represent multiple drainage events. Assuming that lengthy reaches of valley cannot really be flat, and that valley-deeps must have exits somewhere, a valley bottom may have been missed consistently by both geophysics and test drilling at several sites in the central valley and at the exit of the western valley. (See related discussion in Gray, this volume.) This says something about the form of the valley at those places; that is, the floor may resemble a scabland-like topography similar to the klintar of the Maumee-Wabash Trough (discussed in Bleuer, 1989). If so, the existence is implied of intermediate surfaces at about 400 ft (western valley) and 450+ ft (eastern valley).

But alternatively, valley-floor elevations may represent a single, varying profile. The steeper gradient within the western Marion valley and within the Logansport bend (west of DNR test hole 9a; Fig. 3C; also Fig. 5, below) could represent a regional knickpoint related to the valley crossing from limestone to siltstone/shale terrain (Fig. 3) in approaching the Battle Ground lowland. However, this scenario does not explain real, isolated deeps of the Battle Ground lowland and the Mahomet valley (represented only in part by the 70a site; Fig. 3A). Similar apparent deeps are mapped in the Anderson and New Castle valleys, and in the upper Wabash bedrock valley to the south (Gray, 1982). Neither does it explain the flat profile defined by most points in the eastern valley (Fig. 3C).

In addition, valley-floor anomalies can be explained as a result of either of two external events. If deeps are, in fact, isolated, they could be explained either by incision through forebulge or erosion by concentrated sub-ice drainage under great head. If an early, Michigan Basin–centered ice mass repeatedly reached an equilibrium position at the topographic rise onto the limestone terrain of central Indiana, as did late Wisconsinan ice of the Lake Michigan Basin, and if that mass was thick, and of high profile, a forebulge band could have passed through both the present areas of the western deeps and the eastern, flat profile,

---

*Improbable valley-floor elevations include some that must have resulted from poor surface elevation control, including <360 ft at Delphi (Logan, 1920; Horberg, 1945, p. 358) and 300 ft at Oxford (Leverett, 1895), the latter apparently an unlocated well record tied to a nearby railroad elevation. The records include some that may reflect depths of casing driven into shale below the drift, such as the oft-cited 410-ft floor elevation at La Fontaine (Capps, 1910, p. 226; Wayne and Thornbury, 1951, Plate 6).

A 300-ft contour was mapped in Benton County by Horberg (1945, Fig. 4), in Tippecanoe County by Rosenshein (1958, Plate 10), and within the entire Indiana part of the Mahomet valley (Burger and others, 1966) apparently on the basis of the well reported by Leverett. A 200-ft contour was mapped in Tippecanoe County by Maarouf and Melhorn (1975), apparently on the basis of overly deep (hidden-layer problem) IGS seismic-refraction data.

leaving the Logansport area as a hinge line. Such linkage of forebulge and the prerecovery development of deep drainage anomalies has been speculated elsewhere in the Midwest (Frye, 1963; McGinnis, 1968; McGinnis and Heigold, 1973, 1974; Kempton and others, this volume); post-Wisconsinan, ongoing, crustal depression related to just such a forebulge is documented in coastal Maine (Belknap and others, 1987).

A more attractive explanation is erosion by water under ice, which may be possible in general (e.g., Shaw, 1989), or which may relate to catastrophic discharges from beneath ice dams (e.g., Waitt, 1985; Baker and others, 1987). This explanation relates the flat, eastern valley reaches to discharge from beneath the Wilshire ice dam at the St. Marys bend, and relates the western deeps to discharge beneath the West Lebanon or later ice dams downstream from the Logansport bend. In both cases, erosion would have been enhanced by the presence of shale bedrock in the valley floors. This same explanation could be used to explain deeps in the New Castle valley and in the Wabash bedrock valley below Attica or elsewhere in the Midwest (Gray, 1982; McGinnis and Heigold, 1973).

GEOLOGY AND HISTORY OF THE VALLEY CAP AND VALLEY FILL

Regional stratigraphic background

Analysis of the fill of the Lafayette valley system depends on the identification of regionally traceable till sheets above and within the valley fill and identification of the large-scale sedimentary sequences with which the tills are associated. The tills and associated vertical sequences must be traceable on the basis of their own physical properties, which for regional purposes must include downhole geophysical-logging characteristics.

The till sequence in west-central Indiana resembles that documented in Illinois (Fig. 4, Table 3; Bleuer, 1975; Bleuer and others, 1982, 1983; Johnson and others, 1971, 1972; Johnson, 1976; Kempton and others, this volume). However, the fundamental basis of classification is altered here. Most deposits of the Illinois stratigraphic hierarchy had sources in the Superior Province (low garnet/epidote ratio, low bulk magnetic susceptibility), and units of the existing Indiana stratigraphic terminology had sources in the Grenville Province (high garnet/epidote ratio, high bulk magnetic susceptibility) (Wayne, 1963; Bleuer and others, 1983). Because the stratigraphic sequence of northern Indiana records interleaved till sheets of both Grenville and Superior Precambrian source-province mineralogy, basic classification here is based on mineralogic characteristics indicative of these source provinces (Fig. 4). For example, the distinction between Wisconsinan tills of northern and eastern source, tills of the Wedron and Trafalgar Formations, respectively (Fig. 4), is paralleled in older units by assignment of the newly defined West Lebanon Till Member to the Banner Formation and reassignment of the Hillery Till Member to the Jessup Formation.

Units previously defined in Illinois, and now recognized in Indiana in this study, include loam-textured tills of the Vandalia Till Member of the Glasford Formation, and the Hillery Till Member, reassigned here to the Jessup Formation (Fig. 4). Newly

POLARITY CHRONO-ZONES[1]	AMINO-ACID ZONE[2]	MARINE ISOTOPIC STAGE[3]	CHRONO-STRATIGRAPHIC UNIT	LITHOSTRATIGRAPHIC UNIT			
				SUPERIOR PROVINCE	SOURCE MINERALOGY		GRENVILLE PROVINCE
BRUNHES NORMAL — L — — L —	A — B	2	PLEISTOCENE SERIES — WISCONSINAN STAGE	WEDRON FORMATION — SNIDER TILL MEMBER / FAIRGRANGE TILL MEMBER		UPPER TONGUE / LOWER TONGUES	TRAFALGAR FORMATION
		6	ILLINOIAN STAGE	GLASFORD FORMATION — VANDALIA TILL MEMBER		"BUTLERVILLE" TILL MEMBER[4]	JESSUP FORMATION
— L — MATUYAMA REVERSED — WL	C(?) — D	12 | 18 — 22	PRE-ILLINOIAN STAGES	BANNER FORMATION — HARMATTAN TILL MEMBER / WEST LEBANON TILL MEMBER	MAHOMET MEMBER / BLACKFORD MEMBER	BROOKSTON TILL MEMBER[5] / HILLERY TILL MEMBER[6] / WILSHIRE TILL MEMBER	

Figure 4. Lithostratigraphic, chronostratigraphic, polarity chronostratigraphic, and other classifications of units associated with the Lafayette Bedrock Valley System. Key to superior numbers in figure: [1]L = Local normal-polarity zones (Bleuer, 1976, this report; Johnson, 1976); WL = West Lebanon Reversed-Polarity Zone (this report); [2]aminozones of Miller and others (1987); [3]marine isotopic stages of Shackleton and Opdyke (1976); [4]units of western Indiana are not necessarily correlative with Wayne's (1963) type Butlerville Till Member of eastern Indiana; [5]possible equivalent of the Tilton Till Member of Illinois; [6]possible equivalent of the Cloverdale Till Member of western Indiana.

TABLE 3. LITHOLOGIES, TYPE SECTIONS, AND DISTRIBUTION OF NEWLY DEFINED OR REDEFINED STRATIGRAPHIC UNITS OCCURRING WITHIN OR ATOP THE VALLEY FILL OF THE LAFAYETTE BEDROCK VALLEY SYSTEM

Unit	Lithology	Type Section	Distribution
West Lebanon Till Member, Banner Formation	Silty clay loam till	Opossum Run section, Warren Co., Indiana N1/2NE1/4SW1/4,Sec5,T.20N,R.9W	Great Bend of the Wabash (surface exposures); across north-central Indiana (water wells and USGS test holes); Marion valley, Peru segment (pit floors, water wells, and DNR test holes); Marion valley other segments (IDNR test holes)
Blackford Member, Banner and Jessup Formations	Stratified clays, silts (central lacustrine facies); sand to gravel to boulder gravel and tills (eastern and western ice-proximal facies)	DNR test hole 17, Grant Co., Indiana; 10 ft west of Grant-Blackford Co. line; NE1/4NE1/4NE1/4,Sec13,T.26N,R.3E	Marion valley (DNR test holes)
Wilshire Till Member, Jessup Formation	Loam till	DNR test hole 15, Adams Co., Indiana; 4 1/2 mi south of Wilshire, Ohio; NW1/4NW1/4NE1/4,Sec.27,T.26N,R3E	Easternmost Marion valley (DNR test holes)
Brookston Till Member, Jessup Formation	Loam till	DNR test hole 7, Carroll Co., Indiana; NW1/4SW1/4SE1/4,Sec.34,T.26N,R.3E	Battle Ground lowland and Mahomet valley (DNR test holes)
Mahomet Member, Jessup Formation	Sand to coarse gravel	(Reference sections) DNR test hole 1, Benton Co., Indiana; NW1/4SW1/4NW1/4,Sec.31,T.24N,R.9W and DNR test hole 7 (above)	Logansport bend, Battle Ground lowland, and Mahomet valley (DNR test holes)

Note: Units are first described or redefined here.

recognized and defined units (Table 3) include loam-textured tills of the Brookston and Wilshire Till Members (new names) of the Jessup Formation, and the Blackford Member, a glaciolacustrine unit. The latter unit interfingers with both the Wilshire till and the West Lebanon till (described below) and is assigned to *both* the Banner and Jessup Formations. It forms part of the West Lebanon megasequence.

Also newly defined is the unusual West Lebanon Till Member (new name) of the Banner Formation, the oldest known till in Indiana, and that most central to the history of the Marion valley. The West Lebanon tills are reddish, smectitic, silty clay loam- to sandy loam-textured diamicts that contain clasts of red claystones derived from the Jurassic red beds of central Michigan (Bleuer and Moore, 1977; Bleuer and others, 1983). In west-central Indiana, this till lies on proglacial sediments that exhibit reversed remanent magnetism and that contain molluscs exhibiting high amino-acid racemization values (Mill Creek section of Bleuer, 1976; Whites Branch section of Bleuer and others, 1983; Miller and others, 1987). Its deposition therefore is presumed to predate 0.73 Ma, the boundary between the Matuyama Reversed and Brunhes Normal Polarity Chrons (Fig. 4, Table 3).

For present purposes, the till must be assumed to represent a single, albeit complex, glacial event. This may be an oversimplification. The unit is represented by a variety of diamicts in a variety of positions within a West Lebanon megasequence.

The utility of these stratigraphies derives in large measure from the presence of alternating till-beds of differing source composition. Clearly, these markers are expected to be discontinuous laterally. In fact, type areas for components of the Jessup Formation, and most of central Indiana, lie entirely within the realm of eastern-source loam tills. Therefore, lobe-to-lobe correlations for older units are not well established, correlation of some units within the Marion-Mahomet valley-fill (such as the Brookston till discussed below) remain unclear, as do interpretations of some events in valley history.

DNR-Teays and USGS drilling programs

The primary data sources are 21 test holes drilled at sites along the main trunk of the Marion-Mahomet valley (Fig. 5). Sites were chosen with the intent, but not always with the result, of penetrating the thickest valley fill at each location. The 6¾-in.

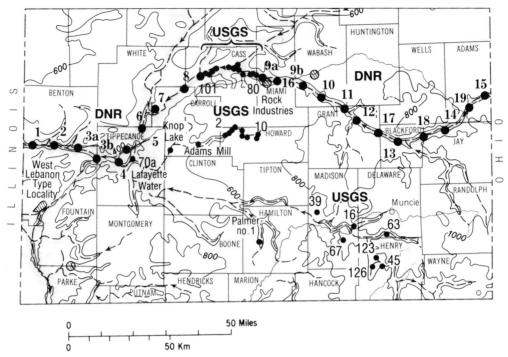

Figure 5. Distribution of DNR, USGS, and other selected test holes in the Lafayette Bedrock Valley System (as defined by the 600-ft contour), and areas of surface exposure of the West Lebanon Till Member (striped pattern) (Refer to Fig. 6 for complete order of USGS holes; base map is adapted from Indiana Geological Survey, 1983.)

diameter mud-rotary test holes ranged in depth from 196 to 433 ft. Samples were collected minimally at 5-ft intervals during drilling; drilling was stopped and mud was circulated at many contact horizons. Samples of granular material were collected in various fine screens, and chip samples of clayey sediments were collected in common french-fry baskets of about ¼-inch mesh, washed in clean water, and air-dried. Collected in this manner, individual cuttings retained their integrity and could be individually hand-cleaned and sorted prior to description and laboratory testing. The inner diameter of the drill string was sufficient to allow gamma-ray logging through the rod before abandonment and plugging. Logging was done by IGS with a Widco Model 1200 gamma-ray logging unit, with 1 11/16-in. diameter tool, at a rate of 15 ft/min, and time constant of 3.

Other test holes referenced here, drilled for groundwater projects of the United States Geological Survey (USGS), are 6¾-in. holes cased with either 2-in. steel or 4-in. PVC. These projects included tests within the Logansport Bend Section and the Peru Segment of the Marion-Mahomet trunk valley in Cass County (Gillies, 1981), and within the upper Wildcat bedrock valley and the Anderson and New Castle valleys (Smith and others, 1985; Lapham, 1981; Arihood, 1982; Arihood and Lapham, 1982; Lapham and Arihood, 1984). Holes in Cass County were partly sampled by M. C. Moore and were gamma-ray logged by the USGS and subsequently also by Bleuer. All deep Wildcat valley tests and some Anderson tests were sampled and logged by Bleuer. Although most are not discussed individually here, additional municipal and private wells were logged and/or sampled by Bleuer in the vicinities of Oxford, Peru, Marion, Geneva, Berne, and West Lafayette (municiple and Purdue University).

Downhole gamma-ray log profiles of these tests provide the tool for integrating large vertical sequences with till stratigraphy, as provided by detailed sample study and laboratory analysis. The profiles provide our first view of the sedimentologic and stratigraphic elements of the fill of the Lafayette Bedrock Valley System (Fig. 6).

Capping units

As much as half of the unconsolidated materials overlying the Marion-Mahomet trunk valley lie above the level of adjacent rock rims of the valley; that is, it caps but does not actually fill the valley. Therefore, stratigraphic identification of valley-capping material is precursory to identification of valley-filling material. In most places, such as across the Marion valley, this material is entirely of Wisconsinan age (Figs. 6, 7, 8). Elsewhere, such as across the Mahomet valley and the Wildcat bedrock valley, the material is Wisconsinan and Illinoian or older (Figs. 7, 8A). The Wisconsinan base is defined in western Indiana by the dated base of till of Fairgrange lithology, in central Indiana by the dated base of till of Trafalgar lithology, and in eastern Indiana by the base of clayey till of Lagro lithology.

In addition, interlobate stratigraphy exemplified by the Wisconsinan deposits provides a gross, but complete, model for stratigraphic relations that might be paralleled within the valley fill. That Wisconsinan model of interleaved deposits of northern and eastern sources is paralleled regionally in the incomplete record of older deposits (Fig. 4) and in the record of the valley fill of the Lafayette system.

Furthermore, the Wisconsinan sequence model illustrates the cutout-inset stratigraphic relations that exist within the middle and upper levels of fill in tributary and trunk valleys. For instance, where the present Wabash River valley parallels or crosses the Marion-Mahomet axis in north-central Indiana, stratigraphy suggests an immediately pre-Wisconsinan Wabash grade similar to that of the present Wabash River (base of USGS 108, Fig. 6B); this grade is possibly equivalent to that defined by the <700-ft rock surface flanking the valley trench up-valley. In the La Fontaine Segment of the central Marion valley, basal Wisconsinan(?) gravel, a major regional aquifer, fills the broad upper valley defined by the <700-ft flanking surfaces (Figure 3D; DNR 10-18, Fig. 6C, D; also Fig. 16 below). Data are insufficient to indicate whether the body is convex upward, as might represent massive, coalesced pro-Trafalgar fan, or whether it is convex downward, as might represent a broad, channel-filling, braided-stream deposit, following or obliquely crossing the trend of the flanking surfaces of the valley. Sharp-based log profiles suggest an erosional base rather than a coarsening-upward proglacial fan progression. Whatever its specific origin, it is technically confined within upper confines of the Lafayette bedrock valley; its base defines the grade of a drainage that apparently opened into the pre-Wisconsinan Wabash described above. And finally, multiple Wisconsinan and pre-Wisconsinan grades to the Wabash appear to be present in the fill of the Wildcat bedrock valley, as discussed below. Cutout and resumed drainage at Wabash grades and other grades, through middle and lower segments of the main valley and its tributaries, is a recurring theme of this discussion.

In contrast, in the narrow eastern segments of the Marion valley, and also in its southern tributaries, Wisconsinan tills define a profound unconformity atop the relatively ancient sediments of the Blackford Member (defined below) (Figs. 7, 8B, C, E, F). Typically, this unconformity is at the level of the flanking rock rims of the valley, indicating that Blackford sediments were repeatedly protected from cross-valley glacial erosion during the Pleistocene. Similarly, in the upper Wildcat bedrock valley (Fig. 8D), the Wisconsinan base lies at about rim level, but is separated from Blackford sediments by a sequence of tills that are separated by evidences of cut-and-fill episodes. These episodes are related to drainage events within the Marion-Mahomet trunk valley and/or to the Wabash bedrock valley. In the Mahomet section (Fig. 8A), the equivalent rim-level cap is the Vandalia Till Member of Illinoian age. Blackford sediments are absent here; that is, a post-Blackford unconformity is interpreted at the valley floor, and an immediately pre-Wisconsinan unconformity occurs near the rock rim. The Wisconsinan base is unrelated to any significant intermediate grade to the Wabash bedrock valley or to the Maumee-Wabash Trough. Thus, the hiatus represented by the great valley-rim unconformity of the Marion Valley Section equates with a post-Blackford time span that is represented by sediments and other unconformities within the western valley sections.

Early concepts of the valley fill

As late as the mid-1970s, concepts of the so-called Teays Valley in Indiana were based on a nebulous sense of location and form, as discussed above, and on virtually no subsurface data within the actual rock confines of the valley. The few drillers' logs available gave little indication of the nature of the fill of the trunk valley, but they gave a somewhat better impression of the lacustrine-clay fill of the upper part of the Anderson and New Castle valleys (Wayne, 1956). For this reason, impressions of the nature of the valley fill have come primarily from studies done down-valley in Illinois and up-valley in Ohio.

In Illinois, paleosols traced across the valley by well records provided a framework for stratigraphic interpretation (Horberg, 1945). The valley fill contains thick, extensive deposits of sand and gravel that make up the Mahomet Sand Member of the Banner Formation (Horberg, 1945; Stephenson, 1967; Kempton and others, this volume). That this material is glacial outwash is clear (Manos, 1961), and its source must have been somewhere east of Illinois.

In contrast, in west-central Ohio the valley fill has been described as mostly "dull blue gray to brown, soft, and highly plastic" clay, commonly more than 200 ft thick and as high as 800 ft in elevation. The clay is associated with lesser amounts of fine sand, "quicksand" or "heaving sand" of local drillers. These materials were interpreted as deposits "laid down in flooded valleys, the coarser sediments originating mainly as outwash from the distant ice sheet, comingling with the finer-grained materials carried into the ponded basin chiefly by headwaters streams." Further, "that the [pre-Illinoian] ice did not reach the area . . . is shown by the absence . . . of ice-laid deposits and of typically coarse outwash deposits" (Norris and Spicer, 1958, p. 226; Goldthwait, this volume).

The background rationale for the present study was given by the knowledge that "somewhere downstream [between west-central Ohio and Illinois, i.e., in Indiana] . . . the lower course of the Teays River was dammed by an early glacier. In that area the lacustrine clays and sands can be expected to give way to glacial deposits that probably include buried outwash sands and gravels that would be productive of ground water" (Norris and Spicer, 1958). In addition, somewhere in Indiana should lie the glacial stratigraphic record that would tie eastern lacustrine sediments, presumably including the Minford Silts of the exposed Teays Valley (Tight, 1903; Bonnett and others, this volume), to the midwestern glacial sequence.

Stratigraphy of the fill of the eastern trunk valley (Marion Valley Section) and the southern tributaries

The Marion Valley Section differs from sections downstream in both fill stratigraphy and morphology. Sequences filling the Marion valley record glacial damming on both the west and

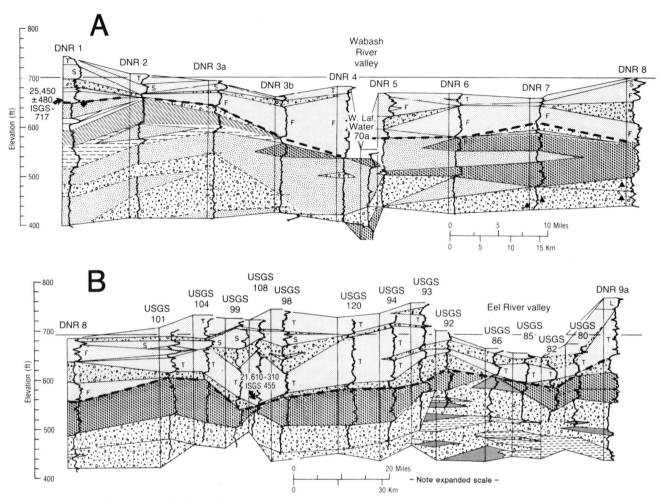

Figure 6. Longitudinal cross sections along parts of the Marion-Mahomet trunk valley showing gamma-ray log traces and stratigraphic units. Refer to Figures 3D and 5 for locations of test holes. A, DNR 1-8; Mahomet Valley Section—Battle Ground Lowland Section. B, DNR 8—USGS Cass County—DNR 9a; Logansport Bend Section. (Note expanded horizontal scale relative to Figs. 6A, C, and D). C, DNR 9a-13; Marion Valley Section, Peru and La Fontaine Segments. D, DNR 13—St. Marys, Ohio; Marion Valley Section, Geneva and St. Marys Bend Segments. Note on gamma-ray log traces: inflections caused by increased steel thickness at drill-rod joints are present at 20-ft spacings (DNR tests only).

east, at the Logansport and St. Marys bends. The sediments form a discrete Marion valley plug. Equivalent sediments do not occur west of the confluence of the Metea valley; whatever sediments may have been there have been cut out and replaced by a younger, inset stratigraphy (Fig. 9).

The Marion valley is filled mostly by sediments of the Blackford Member of the Banner and Jessup Formations. The sediments, including basal gravels, are entirely of mixed, glacially introduced, lithologies (cf. Manos, 1961). The member is defined here on the basis of occurrence within that section (Figs. 4, 9; Table 3). The member includes a western ice-proximal facies, a central lacustrine facies (typified at the type boring, Table 3), and an eastern ice-proximal facies; a lacustrine facies alone is present within the southern tributaries. Parts of the member are assigned to either the Banner Formation or the Jessup Formation based on the lithologic change that takes place within the lacustrine facies; the change in color and clay mineralogy reflects affinities to the West Lebanon and Wilshire Till Members of those formations. Those tills intertongue with Blackford sediments within the western and eastern ice-proximal facies, respectively. The boundary drawn between facies is arbitrary; in fact it is probably diffuse (Fig. 9; drawn generally between DNR 17 and 13, Figs. 5, 6C, D). Those parts of the member associated with the Banner Formation are elements of a West Lebanon megasequnce; those associated with the Jessup Formation are elements of a Wilshire megasequence.

Peru Segment. (Note: Data for this segment are derived from the following test holes: USGS 92, 86, 85, 82, 80; DNR 9a, 16, 9b; for more detailed locations of Cass County tests, compare Figures 3D and 6B.) The Peru Segment of the Marion Valley Section is filled by deposits of the western ice-proximal facies of the Blackford Member (Figs. 4, 6B, 9). Whereas Blackford sedi-

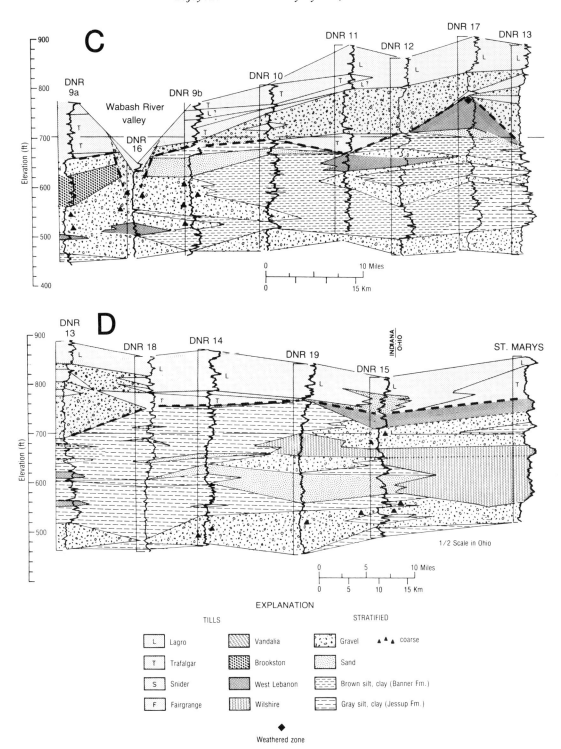

ments are relatively deeply buried in other segments of the Marion valley, they occur at shallow depth in much of the Peru segment, owing to the superposition of late-glacial and postglacial valleys of the Eel and Wabash Rivers (Figs. 1, 5, 6B). Shallow wells and gravel pits penetrate true Marion valley-fill materials within this segment.

This western ice-proximal facies of the Blackford is a complex lithologic association that includes interlayered tills, gravel, sand, silt, and clay. Lentils and tongues of West Lebanon till occur within or atop the unit, mostly in the western part of the facies. The physical characteristics of these tills are much more variable than those of the West Lebanon in its occurrence as in west-central Indiana (Fig. 10A). These units, the "brown" or "red clay" of local drillers, range from reddish brown to brown (7.5YR to 5YR 4-5/4), and from sandy loam to silty clay loam. They commonly are thin, discontinuous, and are interstratified

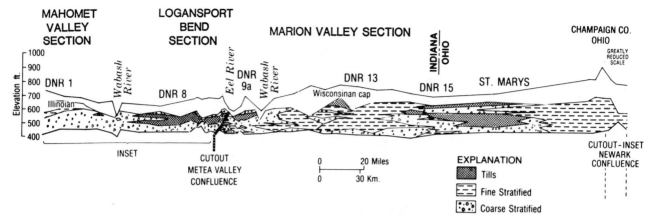

Figure 7. Generalized longitudinal cross section of the Marion-Mahomet trunk valley showing sequences and stratigraphic units, mostly Wisconsinan (undifferentiated), the sub-Wisconsinan valley-rim unconformity, plug and inset sequences of the bona fide valley fill. Refer to Figures 1 and 4 for locations of valley parts and test holes, and to Figures 6 and 9 for stratigraphic details for the Marion Valley Section. (Note the greatly expanded horizontal scale at east end.)

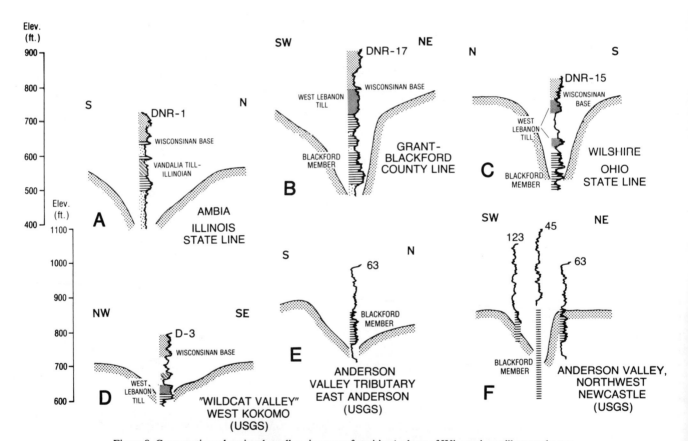

Figure 8. Cross sections showing the valley-rim unconformities (at base of Wisconsinan tills or at the top of Blackford Member) capping sediments of the Marion-Mahomet trunk valley (A through C) and southern tributaries (D through F) and showing evidence of Mahomet or Wabash grades of post-Teays age. Refer to Figure 5 for locations of test holes.

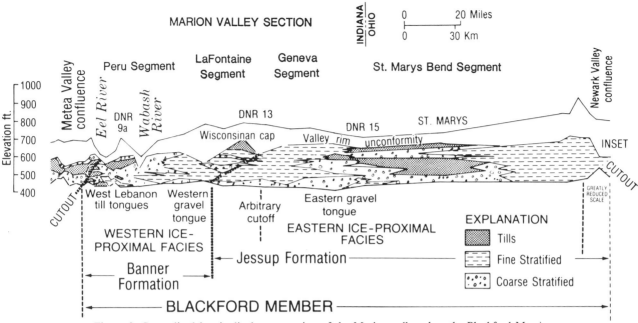

Figure 9. Generalized longitudinal cross section of the Marion valley plug, the Blackford Member, showing relations and nomenclature of the independent, dam-separated, lake deposits of Indiana and Ohio; the till deposits representing the ice dam related to each lake deposit; disconformable cutout and inset relations; and younger stratigraphic units. Refer to Figures 1 and 4 for locations of valley parts and test holes. (Note the greatly reduced horizontal scale at east end.)

with gravels, silts, and clays. Gamma-ray log patterns of West Lebanon materials in this facies vary and include blocky patterns, as well as gradational, coarsening- and fining-upward motifs. The uppermost West Lebanon unit within the Peru segment (USGS 86, Fig. 6B) is more homogeneous, massive, and clay-rich, with a blocky log pattern. (For descriptions of gamma-ray log patterns in glacial materials, see Bleuer, 1986; and Bleuer and Melhorn, 1989.)

In an example of till-facies relations from the only known exposure of Blackford materials, west of Peru and near the south flank of the valley, a thin, loamy till appears to be horizontally bedded, as outlined by thin, slightly carbonate cemented partings (Fig. 11). The base of the unit is irregular and seemingly gradational to the gravel below; the till matrix fills voids between gravel clasts that protrude upward from below. The immediately underlying, coarse gravel is massive and locally contorted, and nearly clast-supported. All deposits rest on poorly exposed, channeled sand and horizontally bedded sandy silt. Massive, reddish, clayey West Lebanon till containing red claystone clasts is found in pit spoil excavated from below this level at the site.

The coarse, stratified sediments appear to be discontinuous tongues, coarsest to the west, where several of the USGS Cass County tests penetrated coarse to bouldery gravel bodies (commonly "broken stone" in drillers' logs; Fig. 6D). These and associated lithologies present an irregular vertical sequence, as evidenced by log patterns. Thinner motifs within the thicker sequence consistently illustrate long-sloping, coarsening-upward profiles, as well as various fining-upward profiles.

Fine-grained, reddish, laminated deposits occur in various positions within the facies (Fig. 6B), including the otherwise coarse, cobbly western part. Laminated beds are more prevalent eastward, where they coalesce with the main body of the Blackford Member.

Most till units within these sections are interpreted as sediment gravity-flow facies of West Lebanon. This interpretation is suggested by the overall associations and geometry, and by a variety of gradational log profiles, which illustrate both coarsening- and fining-upward sequences of sediment progradation. The exposed Peru section (Fig. 11) is representative of these minor, sequences interpreted from downhole logs, that is, of alternating coarsening- and thickening-upward gravel tongues and fining- and thickening-upward till-flow tongues.

Overall, till units interfinger with subaqueous fan gravels, which in turn interfinger lakeward with reddish, laminated clays. Deposits of the lower half of this western facies must represent a variety of types of sediment gravity-flow deposits, ranging from massive matrix-supported to clast-supported tills to coarse, cobbly rubble. As such, these deposits are part of a continuum that probably includes the coarser components of the various gravel tongues. The more massive, blocky-logging capping till (USGS 86) may represent ultimate glacial overriding of the Blackford sediments and subsequent direct bottom meltout of the West Lebanon till.

The red clays are distal lacustrine sediments, whose distribution within the Peru segment was determined by default; that is, they were deposited between aggrading fans or during times of

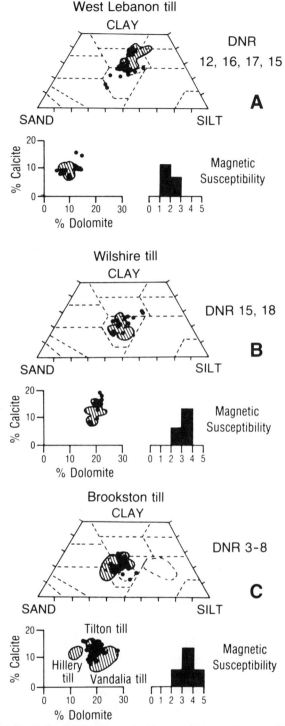

Figure 10. Grain-size, <0.74-mm carbonate, and magnetic susceptibility relations for tills of the Marion-Mahomet valley fill. A, Samples of the West Lebanon Till Member from the central and eastern valley fill as related to samples from surface exposures of the West Lebanon of west-central Indiana (background). B, Samples of the Wilshire Till Member from the eastern valley fill as related to samples of the Brookston till from the central valley fill (background). C, Samples of the Brookston Till Member from the central valley fill as related to samples of the Tilton Till Member from surface exposures of western Illinois (background; data from Johnson and others, 1971).

diminution of glacial discharge. Their occurrence, low in an otherwise most proximal position, illustrates earliest subaqueous fan progradation.

The nature of this association suggests deposition in ice-proximal, subaerial, and subaqueous fan environments associated with a rapidly evolving, near-ice environment, such as those exemplified by tidewater-glacier environments (Powell, 1981; McCabe and others, 1984). Because bottom flows could not have carried the extremely coarse materials far lakeward, the gravels extending eastward from the bend area (USGS 92 to DNR 10) cannot have been derived solely from the ice at a point of valley crossing. Rather, the gravels must represent coalesced subaqueous fans or fan deltas derived from point sources along an extensive ice margin during oblique advance across the valley. (See summary Fig. 14, below).

La Fontaine Segment. (Note: Data for this segment are derived from the following test holes: DNR 10, 11, 12, 17, 13, 18, 14. See also Bleuer and others, this volume, Fig. 2). The La Fontaine Segment of the Marion Valley Section is filled by the central lacustrine facies of the Blackford Member (Figs. 4, 6C, 9; Table 3). This facies is dominated by finely laminated clays, some interlaminated silts, and various larger sand or gravel bodies. Single clay units may exceed 100 ft in thickness (as DNR 12 and 14); however, log patterns show that the clays are heterogeneous, composed of many subtle, internal fining and coarsening sequences. Reddish (7.5YR 5-4/4), smectitic clays dominate the section in the west half of the valley segment and the upper part of the vertical section. They immediately underlie the West Lebanon till cap in the central part of the valley segment. Gray (10YR 5-4/4), less smectitic sediments dominate in the east. Although subtle color variations are common throughout individual borings, an arbitrary vertical boundary separates parts of the Blackford Member assigned to the Banner Formation (west) and to the Jessup Formation (east) (Fig. 9).

A basal sand and gravel body exists throughout the segment. Local gravels are embedded within the clay in the central part of the segment (DNR 11-13). These appear to be isolated laterally and associated with capping West Lebanon tills.

The fine-grained Blackford sediments are considered to be distinctly lacustrine. Internal grain-size variations, indicated by log motifs, and scattered, discrete silt and sand beds, suggest additions to the deeper lake-bottom sequence of bottom-flow sediments related to prograding and retrograding distal fan environments. West Lebanon till-like materials (especially DNR 10 and 12) probably are sediment gravity-flows derived indirectly from ice (tributary-derived turbidity currents) and/or bergstone muds derived from sediment rain from melting of floating ice.

Coarser sediments, and perhaps some of the till-like material, are presumed to represent the entry of outwash and turbidity flows at point sources. These strata are probably in the form of subaqueous fans. Gradational, coarsening-upward sequences embedded within the member represent fan progradation; blocky-logging sands represent mid- and lower-fan channels (Walker, 1984). The point sources are assumed to have been northern tributaries that drained the encroaching West Lebanon ice. One of the primary entry sources must have been rock lowlands and valleys that entered the trunk valley from the north in the Marion area. (See Bruns and others, 1985.) Significantly, the addition of this outwash, and ultimately the capping (basal?) tills, came late in the valley-fill sequence; that is, although the materials may have bypassed their way into the valley through a gap in the north wall, they represent an overall progradational sequence related to the advance of West Lebanon ice, obliquely across the trunk valley. (See summary Figure 14 below.)

Geneva Segment. (Note: Data for this segment are derived from the following test holes: DNR 18, 14, 19, 15. See also Bleuer and others, this volume, Fig. 3.) The fill of the western part of the Geneva Segment of the Marion Valley Section resembles that of the La Fontaine segment, except for the coarser stratified sediments interfingering into it from the east (Figs. 6D, 9). The lowest coarse sediments extend farthest west; higher in the section, finer sediments overstep eastward, the La Fontaine-like, glaciolacustrine facies of the Blackford grading into an eastern ice-proximal facies.

The eastern part of the segment contains the eastern ice-proximal facies of the Blackford Member (Figs. 7, 9). It is a complex of tills, sands and gravels, silts, and clays similar in nature and geometry to the western ice-proximal facies that fills the Peru segment. The facies includes a till unit, the Wilshire Till Member of the Jessup Formation, defined on the basis of this occurrence (Table 3, Fig. 6D); the till is not presently known outside the confines of the valley. The Wilshire is a gray (10YR 5/4 tending toward 7.5YR), loam till of eastern-source mineralogy (Fig. 10B).

In contrast to its western equivalent, the eastern ice-proximal facies contains relatively little coarse gravel, except in the case of thick, bouldery, limestone-gravel units at various levels of DNR 15. The elements appear to be complexly interstratified, where numerous fine-grained beds, even near the base, are interstratified with boundary gravels; the log motif is one of repeated coarsening-upward cycles. Traces of dark-colored, organic loam, calcareous but with traces of secondary carbonate enrichment, occur at about 575 ft elevation (240-ft depth) in DNR 15, and a "green sand" is reported at about 580 ft elevation in the St. Marys record. The deep gravels are cemented with calcium carbonate.

This eastern facies includes the repetitive sequence of relatively thick, gray till and coarse, stratified units interpreted from drilling records and gamma-ray logs from near St. Marys, Ohio (supplied by S. E. Norris, and R. R. Pavey; Fig. 12). The coarse, complex sequence of DNR 15 is assumed to be the western equivalent of the entire package at St. Marys. (The correlation of the Wilshire to St. Marys, Fig. 6D, is based solely on drillers' description, position, and log character.) The relation illustrates a westward thinning of tills and replacement with coarse, cobbly outwash; this is a subglacial-proglacial transition. Westward is a complex, interlayered sequence of stratified deposits transitional to the lacustrine deposits; this is the proximal-distal, subaqueous fan transition. Similarly, the St. Marys sequence is *assumed* to

Figure 11. Two photos showing the nature of the West Lebanon sediment-gravity-flow facies: loam to sandy loam, faintly laminated mudflow (2) atop and partly gradational to massive, contorted, matrix-supported, cobbly gravel, and/or sand debris-flow (3), and silty, fine sand, and silt channel fills (4), beneath pit floor (1), which underlies Wisconsinan bar gravels of what has been termed the "Maumee flood" (Fraser and Bleuer, 1988). The pit floor is the local valley-rim unconformity. Arrows indicate corresponding stone in the top and bottom photos. Scale is in inches and centimeters. (Rock Materials pit, sec. 30, T. 27 N., R. 4 E.; location is atop a minor tributary on south edge of the Peru Segment).

grade eastward, through another coarse, stratified sequence, into the lake clays of west-central Ohio (Fig. 12).

The overall geometry of the eastern sediment association, with an eastward overstepping by fine-grained sediments, at first would appear to be one of a retrograding subaqueous fan. However, the basal gravel tongue (Figs. 6D, 7, 9) is best interpreted as a valley-head subaerial outwash fan, deposited from the Wilshire ice prior to damming of the valley downstream. The coarse gravels of this unit apparently are continuous as far downvalley as DNR 10 of the La Fontaine segment. The bouldery units in DNR 15 may represent either ice-edge tunnel discharge or embedded chutes on local fans (McCabe and others, 1984). The association with finer sediments may represent local blockage of the trunk by outwash entering from the north, via the Berne tributary valley in the vicinity of DNR 19. The middle sand is interpreted as a distal deltaic complex, related to the discharge of a meltwater-fed tributary valley into the then-dammed Marion valley from that Berne tributary.

The Wilshire probably represents englacial meltout deposition, where it is thickest and represented on logs as a stacked, blocky motif (as at St. Marys); it probably represents sediment gravity-flow deposition where it is thin and associated with a complex vertical sequence (DNR 15), or where characterized by diffusely varying log character, and presumably thinly interbedded with sandy sediments (DNR 19). As in other sections, possibly the deep, cemented carbonate represents groundwater circulation in paleolandscapes developed at levels within the deep gravels, as the organic silt and "green sand" may be paleosols.

Southern tributaries. The southern tributaries of the Marion-Mahomet trunk valley are the Anderson and New Castle Valley Sections, the Wildcat Bedrock Valley Section, and the broad Frankfort Lowland Section.

Within the upper Wildcat bedrock valley, the Wisconsinan base forms a valley-rim unconformity like that of the Marion valley (Fig. 8D), but the unconformity surface and older, till-defined surfaces decrease markedly in elevation in a down valley direction toward the Maumee-Wabash Trough (Fig. 13). The Wisconsinan sequence in the Wildcat bedrock valley is marked by a well-defined set of three tills (Fig. 12) overlying finitely dated organic silt and gleyed paleosol. Underlying the Wisconsinan deposits are at least two till units overlying basal sediments of the Blackford Member. The latter includes alluvial and lacustrine sediments capped by West Lebanon till.

The overall sequence filling the Wildcat bedrock valley suggests that several post-Blackford, pre-Wisconsinan levels may have been graded to intermediate drainageways in the trunk valley. This suggests that many possible drainage courses, exiting various routes from the Lafayette hub, may have succeeded initial drainage through the trunk valley.

The Blackford Member itself consists of eastward-rising and eastward-pinching tongues of West Lebanon till that underlie gravel and overlie stratified clay and gravel (Fig. 13). This sequence records a simpler version of the sequences associated with the blockage of the Marion valley. Lake clays were deposited within an independent, glacially blocked valley, followed by glacial flow up-valley and by the deposition of the West Lebanon till as subaqueous flows and basal tills. The presence of interfingered outwash (alluvium?) and lake clay over a vertical range of over 100 ft suggests a progradation of clays, eastward and upward, over valley-head–derived alluvial or deltaic materials in a gradually deepening lake. Yet the dearth of coarser ice-proximal deposits between clay and till suggests a relatively rapid, continuous override.

The Anderson and New Castle valleys are filled with brownish, soft, smooth lake clays to an elevation above 800 ft (Figs. 8E, F), again the level of the local valley rim, and now the level of a valley-rim unconformity. All these sediments are assumed to be remnants of a once much more extensive distribution of the Blackford Member within the drainage basins of the southern tributaries.

Were the southern tributaries and the Marion trunk valley coexistent prior to filling? Stratigraphic evidence suggests that they were, and that the fill of upper reaches of tributary valleys and of the central trunk valley consist of lacustrine deposits of similar age and origin. However, present elevations of the heads of those valleys (the New Castle valley–Whitewater valley divide near Connersville and the Noblesville valley or valleys near Westfield) are below the level of the fine-grained sediments filling them, presenting a further problem.

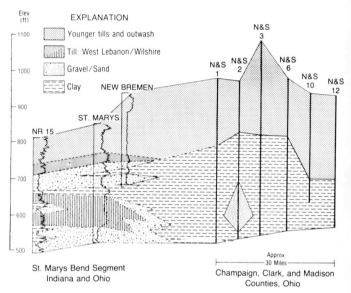

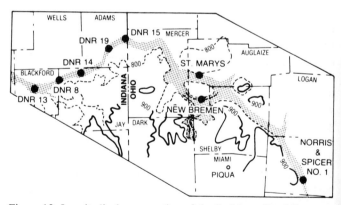

Figure 12. Longitudinal cross section of the St. Marys Bend Segment, connecting Indiana-DNR data (this chapter) and Ohio-USGS data (Norris and Spicer, 1958, Plate 8) through points at St. Marys and New Bremen, showing relations of units in the valley fill. Base map is adapted from Gray (1982) and Cummins (1959). Ohio (USGS) gamma-ray logs were made available by S. E. Norris and R. R. Pavey.

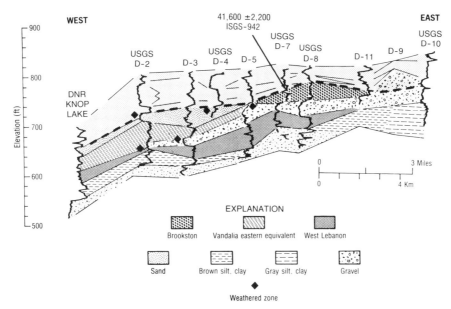

Figure 13. Longitudinal cross section of the Wildcat Bedrock Valley Section, extending westward into the mergence of the Frankfort Lowland and Battle Ground Lowland Sections, showing gamma-ray log traces and stratigraphic units.

Interpretations of Lakes Marion, Wildcat, and Anderson and the sequence of glacial advance

Although long and sinuous, the trunk valley and its southern tributaries are narrow and of relatively small volume. These valleys were sediment traps capable of recording sediment influx related to the most subtle events within the respective basins. Surely, then, the gross geometry of the valley-fill sediments records the massive externally imposed events of glacial advance and drainage.

The stratigraphy outlined above suggests that the valley-fill sediments of the West Lebanon and Wilshire megasequences were deposited during a single episode, or at the most, during a few closely spaced episodes of glacial blockage. Blockage of the Marion valley appears to have occurred independently, but more or less contemporaneously, at both of that valley's northern bends, and lakes formed upstream from those dams. Ultimately, the mouths of the southern tributary sections were blocked as well, creating separate lakes within each (Fig. 14). A variety of glaciofluvial and glaciolacustrine deposits, the Blackford Member, filled these dammed valleys. (Although disjunct with respect to the type sediments within the Marion valley, these deposits are considered elements of the Blackford on the basis of stratigraphic position.)

Lakes Marion, Wildcat, and Anderson are defined here as the water bodies dammed within the respective rock-walled valleys known by those names (Figs. 1, 14). The water levels ultimately reached within each of these lakes may have been regulated by the bottom elevations of any of several valleys or sags opening southward (Fig. 14), although the existence of any of these at the necessary time is speculative, and possible isostatic effects must be ignored. These possible outlets include levels as low as 400 ft (Attica cutoff), 700 ft (Linden and Noblesville valleys), and 800 ft (southwestward trenching sag from the bend at Hartford City, and the New Castle valley–Whitewater valley connection at Connersville).

Assuming that the lowest till fill at St. Marys and Wilshire represents the ice source of the basal, coarse, probably subaerial outwash to the west, then incursion of Wilshire ice into the valley at the St. Marys bend must have immediately preceded any significant damming at the Logansport bend (Fig. 14, stage 1). Interstratified basal clays and boulders at Wilshire (DNR 15) suggest that coarse outwash initially may have come not from the bend proper, but via the Berne tributary, with an outwash fan puddling a small lake in the trunk valley upstream from DNR 19 (Figs. 1, 6D, 9; Bruns and others, 1985).

The Wilshire ice at the St. Marys bend completely dammed the valley upstream, creating a lake there, as discussed by Norris and Spicer (1958); presumably this lake was classic Lake Tight (Hoyer, 1976, 1983), which at some stage reached elevations of 860 ft, and perhaps even 910 ft (Hoyer, 1976). Total damming of the rock valley in the St. Marys area would have created a lake about that deep, depending on the nature of outflow through southbound channels in the Piqua area (Fig. 12).

West Lebanon ice, advancing from due north (from across central Michigan) would have first encountered the Tippecanoe and Metea tributaries (Fig. 14, stage 2). Concentration of West Lebanon meltwater and outwash materials within these valleys would have been the first direct glacial influence on the central Lafayette system, and periodic damming, either by hydraulic effects or by direct deposition, would have influenced deposition in the main trunk valley. This event may be responsible for the lowest clays found in the western ice-proximal facies.

A sedimentologic record of the actual dam probably does not exist; it would have been in the bend proper, within the Metea confluence, the site of subsequent cutout and inset of younger strata. From that dam of West Lebanon ice, however, was derived a variety of sediment-gravity flows in association with morainal bank deposits, coalesced subaqueous-fan gravels, and intrafan lake muds. Till flows, which occur low in the valley and appear to lack continuity, may indicate independent, repetitive events related to the wide ice-front melting and calving along the long north valley wall.

Although the Logansport bend is logically the first spot to assume glacial encroachment into the central valley (Fig. 14, stage 3), an ice front ultimately would have obliquely fronted the

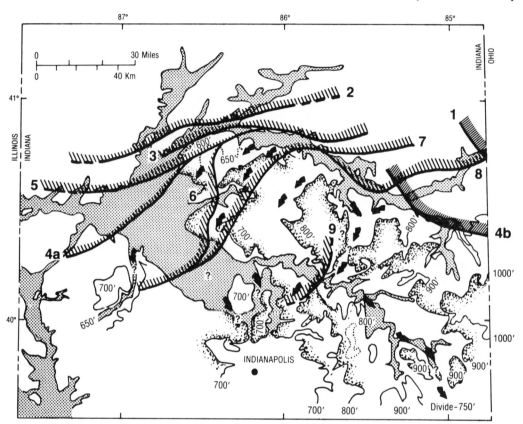

Figure 14. Interpreted stages of blockage of the Marion Valley and Frankfort Lowland Sections and their tributaries by advancing eastern-source (Wilshire) and northern-source (West Lebanon) ice sheets (fine- and coarse-hachured, respectively); possible overflow routes suggested by arrows across generalized contours on upland surface. Base map is adapted from Gray (1982). Refer to Figure 1 for locations mentioned below. Stage 1: Earliest Wilshire ice blocks the St. Marys bend; in-valley deposition of lowest Wilshire till and outwash spreads *down*-valley in Indiana, deposition of the lake sequence begins *up*-valley in Ohio. Stage 2. Outwash from West Lebanon ice spreads down the Tippecanoe bedrock valley and Metea valley at the Metea-Marion confluence; ice blocks the Logansport bend; deposition of subaqueous fan, sediment-gravity flow, and interfingered fine lake deposits (reddish clays) begins to spread *up* valley into Lake Marion. Stage 3: West Lebanon ice lies astride north valley wall of the entire Peru segment, feeding coalescing proximal fans and fan deltas composed of gravels and finer grained gravity flow sediments; background sedimentation of lake muds continues to fill the ice- and outwash-blocked valley; possible periodic drainage open downstream. Lake Marion overflows across the upland south of the bend, transporting gravel, locally rich in black shale (from that upland), into the Wildcat bedrock valley. Stage 4a, b: West Lebanon ice blocks the Logansport bend and Frankfort lowland. Lake muds of Blackford Member begin to accumulate in Lake Wildcat and Lake Marion. Wilshire ice reaches southwesternmost extent atop its fan sediments. Stage 5: Ice retreats from, or ice dam is breached at the Battle Ground-Mahomet confluence, allowing drainage of the lake in the Frankfort lowland and Wildcat bedrock valley; organic soil develops atop red lake clays at the Knop Lake location. Stages 6, 7, 8: West Lebanon ice rides across the Marion valley, forces the level of Lake Marion to exceed 800 ft, and deposits a till cap across a valley already nearly full of Blackford lake sediments. Stage 9. West Lebanon ice dams the mouth of the Anderson valley and forces the lake level of Lake Anderson to exceed 800 ft.

30-mi-long Peru segment (USGS 92-DNR 9a, b) (Fig. 14, stages 3–6). The eastward-thinning gravel tongue of the upper half of the sequence may represent coalesced subaqueous *and subaerial* outwash fans derived from ice that stood precipitously along the north valley wall of the entire Peru segment. The surface level of the earliest Marion lake(s) would have ranged from 550 to 600 ft in elevation, as determined by the configuration of the rock surface on the south side of the Logansport bend. The level of the top of the gravel tongue, ~600 ft, corresponds to the southern rim level, a temporary spillway above which no true lake fill could accumulate (Fig. 14, stages 3, 4).

Concurrently in the Marion area (DNR 11, 12 and new north Marion well-field tests), gravel outwash and/or subaqueous fans built into the valley from drainage entering through the bedrock sag that extends from Huntington to Marion (Fig. 5; Bruns and others, 1982). And in the St. Marys bend area, sandy deltaic and finer prodeltaic deposits, and finally coarse subaqueous outwash fan deposits, entered the valley from the east and north, from St. Marys and from the Berne tributary (DNR 14, 19, 15). This was followed by a westward override by Wilshire ice (Fig. 14, stages 1-4b).

Coarse outwash bodies throughout the Marion valley have been cited as evidence for proximal, mostly subaqueous fan deposition within the valley. However, it is possible that some of these might evidence periods of large-scale lake drainage, stranding of floating ice, and finally development of subaerial channels. If the West Lebanon ice front resided at the Logansport bend during most of Blackford deposition, lake-head and thin ice could have combined to permit periodic uplift of the ice terminus and the catastrophic discharge from beneath this ice in a small-scale equivalent of the events described by Waitt (1985). Indeed, the lakes in the southern tributaries probably drained at least once (discussed below). Water depths critical to both subglacial drainage and to advance across the lake basin (discussed below) could have been controlled by the level of the discharge rate over the south rim (600- to 650-ft contours) and by the elevation of the sediment within the valley.

Once totally obstructing the valley, West Lebanon ice would have raised the level of the backed-up Lake Marion as the ice advanced eastward across progressively higher upland rims. When ice reached the head of the La Fontaine segment, the valley was blocked head-on for the first time, and the lake level would have progressively risen to 700 ft and more (Fig. 14, stage 7). Advance of the ice could have raised this level to as much as 800 ft by the time the ice stood across the Geneva segment (Fig. 14, stages 8-9).

Probably the most important aspect of the Blackford valley-fill sediments deposited ahead of this advance is that the main body of West Lebanon till overlies the sediments and that its contact rises eastward. That is, deposition of the West Lebanon followed and capped Blackford lake sediments within an already full valley. The major occurrences of West Lebanon within eastern valley-fill sediments lie at about the elevation of the rim of the rock valley (~700 ft; DNR 17, 15). In addition, fill of the La Fontaine segment contains relatively little outwash, except that introduced through northern tributaries; fill of the Geneva segment contains neither the red West Lebanon till, nor even red clays derived from it. Thus, fill stratigraphy suggests that West Lebanon ice was nowhere near eastern valley segments during most of the time of lake deposition (Fig. 14, stages 4–6a).

This suggests that lacustrine sedimentation was rapid with respect to marginal ice advance; in fact, the valley could have been virtually filled during the period in which lake levels fluctuated within the 550- to 600-ft range (determined by the rock configuration at the Logansport bend; stages 2, 3, Fig. 14) and during the time that coarse outwash and/or deltaic sediments were being debouched from northern tributaries in the Marion and Berne areas.

The rapidity of sedimentation here cannot be determined; however, an analogy might be the accumulation rate for muds at the front of a tidewater glacier, which Powell (1983) has documented as more than 4 m/yr. Assuming that the interpreted expanse of glacier front residing on the north rim of the narrow Marion valley might compensate for possible density differences of the sediments, and for the obvious differences between tidal and lacustrine environments, this means that 200 ft of Blackford clays could have been deposited in only 15 yr. In that time, the front of an ice sheet moving 200 m/yr would have advanced less than a mile. No wonder that evidence of near-ice deposition, such as elements of the western ice-proximal facies, diminish rapidly eastward, and that basal West Lebanon tills cap a full valley. (The rapid rate of filling suggested here contrasts with a much slower rate far upstream of the blockage of the St. Marys bend: based on varve counts in the Minford deposits of southern Ohio, Hoyer (1976) has suggested Lake Tight's duration to be 8,000 to 10,000 yr. Inasmuch as the Marion section was forever plugged after this time, that duration need not require *ice* damming of the lake. Rather, it records time before breeching of a southern outlet, and the beginning of classic "deep stage" incision in Ohio.).

Blockage of the broad outlet of the Frankfort lowland would have come somewhat later than the blockage of the Marion valley at Logansport (Fig. 14, stages 6-9). Basal black shale rubble (USGS D-7), whose source is several miles north of the Wildcat valley, likely was transported into the still-open Wildcat by early outbreaks of flood flow from Lake Marion. Initial blockage of the Frankfort lowland and of the Wildcat bedrock valley was not absolute. The presence of a thin organic silt at the till-clay contact at Knop Lake (Figs. 5, 14) suggests at least one stage of lake emptying prior to glacial override. Coarse outwash or ice-proximal subaqueous sediments, similar to those at the Logansport bend, undoubtedly were associated with the West Lebanon ice front during the closing of the mouth of the Frankfort lowland. Either the sediments have been removed subsequently, or they have not yet been found.

Lake Wildcat ultimately rose from less than 600 ft to about 700 ft, owing to the ice blocking first the mouth of the Frankfort lowland and, finally, bridging the >700-ft uplands flanking the mouth of the Wildcat bedrock valley (Fig. 14, West Lebanon positions 4–6a). This lake would have extended into the Anderson valley beyond Anderson. The Linden and Noblesville valleys, if in existence in present form, would have stabilized the lake at ~650-ft and ~700-ft elevations or at some intermediate levels balanced by overflows from Lake Marion (during the time of the rise of Lake Marion from ~650 ft to just under 800 ft).

Only after West Lebanon ice blocked the Noblesville-Anderson confluence west of Tipton could waters in Lake Anderson have exceeded 700 ft. While there is presently no evidence for this blockage, it must have occurred, because the upper Noblesville valley *was* blocked. Samples and gamma-ray log of a petroleum test near Westfield, just north of Noblesville (Eldon Palmer no. 1, Wabash Resources, Inc.; Fig. 5), verify that the valley is now plugged by a single, thick till of West Lebanon lithology. Ice within the Anderson valley would have had to pass well beyond Anderson to effect closure of the 800- and 850-ft contour levels across the rock uplands. (See Gray, 1982.)

The connection between the bedrock New Castle valley and the modern Whitewater River valley at ~750-ft elevation remains enigmatic, because fine-grained Blackford sediments fill the New Castle valley to its present valley rim, >800-ft in elevation. If the connection existed during Blackford time, outflow from it must have been impeded by a drift or ice dam related to an Ohio ice lobe. Such a dam could be related to Wilshire, or older, ice. In fact, magnetically reversed lacustrine sediments within the Whitewater basin (Handley Farm section, Bleuer, 1976) could be correlative with the Blackford Member. Alternatively, headward erosion of the Whitewater may have produced the connection in post-Blackford time.

Stratigraphy of the fill of the central and western trunk valley

As subsequent events continued to shape the Lafayette Bedrock Valley System, the sediments of the Marion valley plug were left hanging above a reexcavated Metea valley at the Marion-Metea confluence in central Indiana. The deposits filling the Logansport Bend, Battle Ground Lowland, and Mahomet Valley Sections, the Marion-Mahomet trunk valley downstream from the Metea confluence, are younger than and are wholly inset into sediments plugging the Marion valley (Fig. 7).

The foregoing interpretation rests not only on recognition of the lateral relations of units within the valley fill, but also on identification of the base of the Wisconsinan Stage as an aid in knowing what actually is valley fill. Wisconsinan sediments are defined by the dated base of the Trafalgar Formation in USGS 108 and by correlations therefrom. The Trafalgar is readily recognized by lithology, logging characteristics, and stratigraphic position. Westward this position is assumed by another unit, correlated with less assurance with the Fairgrange Till Member of Lake Michigan Lobe source. That correlation is extended from the dated base of the Fairgrange in DNR 1, through Lafayette area test holes and valley-wall exposures, and thence eastward. The extent of presumed Fairgrange closely matches that of the Snider Till Member, a marker unit more positively traceable by logs into the USGS Cass County test holes.

Logansport Bend and Battle Ground Lowland Sections.
(Note: Data for this section are derived from the following test holes, east-to-west: USGS 93, 99, 120, 98, 108, 99, 104, 101; DNR 8, 7, 6, 5, West Lafayette Water Co. 70a, DNR 4; for more detailed locations of Cass County tests, compare Figs. 3D and 6B.) The primary component of true valley fill of the Logansport Bend and Battle Ground Lowland Sections is the Brookston Till Member of the Jessup Formation, which is defined here on the basis of that occurrence (Table 3). The Brookston till and the Mahomet Member (discussed below) are inset within the west cutout of the Blackford Member that occurs in the short distance between USGS 92 and 93 (Figs. 5, 6B). The Brookston till is a gray (10YR 5/2) loam, except at West Lafayette where the material of the test 70a, sampled by power auger, is considerably more silty (Gardner, 1979). It is characterized by high carbonate content, relatively high calcite/dolomite ratio, and very high magnetic susceptibility values (Fig. 10C). The Brookston is a single unit more than 50 ft thick in Cass County (USGS 101-93), but farther west it splits into at least two units (Figs. 6A, 7). The Brookston does not resemble most Trafalgar tills or presumed Illinoian tills presently recognized in the subsurface of central Indiana (Bleuer and others, 1983). The Brookston more resembles the Tilton Till Member (pre-Illinoian) in Illinois and the uppermost Jessup till (so-called Illinoian) of southwestern Indiana.

The Brookston till, which overlies the Blackford Member and the included West Lebanon till at the Logansport bend (Fig. 6B), is the only significant glacigenic unit within the sediment mass of the western valley. However, some remnant West Lebanon may remain. New data from a Purdue University test at West Lafayette suggest that a sub-400-ft depression there is filled by a sandy reddish till (the West Lebanon?), which is separated from the overlying Mahomet sediments by a thin marl bed at the 400-ft level. Another boring set in south Lafayette (Lafayette Country Club) suggests that West Lebanon is plastered against the south valley wall.

The stratified sediments of these valley sections are considered correlatives of the single, massive Mahomet Member of the Mahomet section to the west (Fig. 4). (DNR tests 1 and 7 are considered primary reference sections [Table 3; Figs. 5, 6A].) Within the Logansport bend (USGS tests 101-93), the Mahomet Member lies beneath Brookston till and rests on rock. The unit contains fine, gravelly sand to coarse gravel, or gravel with "broken stone," a driller's description identical to that interpreted as bouldery fan gravels within the Marion section. However, most sections are of coarse sand and fine to medium gravel that may include alternating clay interbeds. Stratified sediments immediately below the Brookston are consistently finer than those lower in the section. Within the Battle Ground lowland (USGS tests through DNR 6), the Mahomet splits between tongues of the Brookston. The upper unit is sand, and the lower unit is mostly coarse gravel (Fig. 6A).

Mahomet Valley Section. (Note: Data for this section are derived primarily from the following test holes, east-to-west: DNR 3b, 3a, 2, 1. See also Bleuer and others, this volume, Fig. 3.) The fill of the Mahomet Valley Section consists of massive deposits of predominantly sand and medium to fine gravel of the Mahomet Member of the Jessup Formation (Fig. 4). Because the unit appears to be part of a Brookston megasequence and appears to be dominantly of eastern source, it is redefined here as part of the Jessup Formation. No intermediate marker beds are obvious, and no consistent lithologic variations can be generalized. Cemented zones are present at varying levels. The unit's surface reaches nearly 600 ft in elevation (DNR 3a), but westward (DNR 1) that surface drops and is capped by laminated clays; paleosols at and slightly below this surface in DNR 1 occur at about the 500-ft elevation, a level about the same as that of flanking, intermediate rock surfaces suggested by gravity profiling (Fig. 3A).

The Mahomet Member can be interpreted as a single outwash body related to Brookston ice, that is, part of a Brookston megasequence, or it can be interpreted as superimposed outwash bodies, separated by buried erosion surfaces related to drainage either to the west or south (via Attica). Clearly, owing to the great width of the Mahomet valley (Fig. 3), the line of section is far from adequate, and no hypothesis about the origins of Mahomet sediments can be complete or wholly substantiable.

Interpretations of events and drainage systems through time—a string of assumptions

How many drainage systems did the Lafayette Bedrock Valley System hold, at what levels, in what sequence, and going where? High-level surfaces flanking the middle and lower trunk valley suggest some form of pre-classic Teays drainage, extending into the lower valley route from the shale lowlands of northeastern Indiana (Fig. 3D, and frontispiece; drainage stage I, Fig. 15). A main bench and deeper holes in the lower valley do not necessarily suggest multiple drainage events or stages. Certainly, many old questions regarding correlations and the origins of Teays and so-called deep stage drainage remain unanswered and are perhaps moot.

For purposes here, the history of drainage within the system necessarily begins with an *assumed* major continuous valley system already in existence sometime prior to 0.7 Ma, owing to whatever mechanisms (Gray, and Melhorn and Kempton, this volume). This through-flowing drainage, the classic Teays-Mahomet, entered Indiana from the east and passed out to the west (i.e., it did not pass down an ancestral Wabash River; drainage stages II–IIIa, Fig. 15).

The stratigraphic reference for the end of this period of Teays-Mahomet drainage is the West Lebanon till. If the West Lebanon megasequence represents a single glacial event, its presence in the Marion valley, and possibly as residual bottom deposits in the Mahomet valley, establishes the earlier continuity of those sections. The West Lebanon ice advance across those valleys created and filled Lake Marion in Indiana, and the Wilshire ice created a separate lake upstream in western Ohio, probably the same lake in which the Minford Silts were deposited far upstream (drainage stage IIIb, Fig. 15).

The extent of any equivalent filling of the trunk valley below Logansport (Logansport bend, Battle Ground lowland, Mahomet valley) is unknown, and it is presumed to have been removed during an erosional episode that produced a combined Metea-Mahomet valley (drainage stage IVa, Fig. 15). (In very broad concept, this episode is equivalent to the deep stage of Ohio—it produced a valley cutting through Teays-age valley-fill sediments.) Remnants of the original fill below the Logansport bend may include parts of the basal gravel of the Mahomet valley, the sub-400-ft level marl and till sequence at West Lafayette, and a red till in tributaries of the Mahomet Bedrock Valley in Illinois (Kempton and others, this volume).

Interpretations of drainage evolution after blockage of the Marion valley are more hypothetical. They are based on the following string of facts and assumptions (refer to Figs. 4 and 7): (1) The westward-pinching tongues of Brookston till of the Battle Ground lowland, which interfinger with the Mahomet, are distinctly older than the Vandalia till rather than simply eastern-source equivalents of it. This is so because a paleosol (magnitude unknown) exists high within the member, below Vandalia till (Illinoian) in DNR 1 (Fig. 6A). (2) This Brookston-Mahomet sequence is younger than the West Lebanon till (by superposition at Logansport and by apparent cutout to the west). (3) The Harmattan, Hillery, and Tilton tills are also older than the Vandalia and younger than the West Lebanon (by position in Illinois and in the vicinity of the West Lebanon type area in west-central Indiana (Fig. 5). And, (4) the Harmattan and West Lebanon tills are not merely representative of differing flow-path facies or differing lobes of a single glaciation; they exhibit normal and reversed remanent magnetism, respectively (Johnson, 1976; Bleuer, 1976), a condition most readily explained by deposition during different glacial events. Therefore, the placement of the Brookston-Mahomet sequence within a regional stratigraphic framework would seem simple, except for the fact that: (5) in western Indiana, no tills older than Vandalia are associated with the Mahomet sediments, and in Illinois, the position of the Harmattan with respect to the Mahomet Sand Member is not clear. However, multiple tills interfingered with Mahomet materials along valley sides and tributaries in Illinois similarly suggest a record of multiple events (Kempton and others, this volume).

In a most simplistic model, an incised-to-rock, Metea-Mahomet drainage (stage IVa) postdated Marion blockage. The subsequent alteration of this drainage must involve regionally recognizable till units related to northern and eastern glacial lobes. In this model, the Brookston-Mahomet sequence within the trunk valley in Indiana is the up-ice, rock-stratigraphic representation of all or part of the Tilton and Hillery till sequence that lies atop the trunk valley in Illinois. Most of the present valley fill was emplaced as the Brookston (Tilton?) till advanced into the valley in Indiana (stage VIb, Fig. 15). (If Harmattan till occurs

Drainage stage	Drainage System	Exit elevation	Valley-fill stratigraphy	Valley filled	Marine isotopic stage
IX	Modern Wabash via Maumee-Wabash Trough into lower Wabash via Independence sill	490 feet (sill level)	Modern atmosphere	Maumee-Wabash Trough	1
VIIIb			Trafalgar Formation, Fairgrange and Snider Till Members Wedron Formation	Tippecanoe, Metea, middle and upper Wabash valleys	2
VIIIa	Tippecanoe and Metea into Wabash via Independence sill or Attica cutoff	490 feet (sill level)			
VIIb			Vandalia Till Member Glasford Formation	Mahomet valley (final cap)	6
VIIa	Mahomet	?			
VIb			Mahomet Member Brookston Till Member Jessup Formation	Mahomet valley Logansport bend, Battle Ground lowland	12-18
VIa	Metea-Mahomet	?			
Vb			Hillery Till Member(?) Jessup Formation	Attica cutoff	?
Va	Metea-Wabash ? via Attica cutoff or Danville valley	<400 feet			
IVb			Harmattan Till Member Banner Formation	Metea and Mahomet valleys	19-21
IVa	Metea-Mahomet earliest "deep stage" equivalent	<400 feet	Mahomet Member Jessup Formation		
			(Valley incision)		
IIIb			West Lebanon Till Member Banner Formation	Marion valley (final plug)	22
IIIa	Marion-Mahomet the so-called Teays (Teays III[2])	350-400 feet	Wilshire Till Member Jessup Formation Blackford Member Banner and Jessup Fms.		
	Teays III[2]	?	(Valley incision)		?
II	Teays II[2]	?	(Valley established)		?
I	Teays I[2]	?	(Northward, westward, southward drainage ?)		?

[1] Marine isotopic stages of Shackleton and Opdyke (1976) relate to age of fill
[2] Valley-development stages of Gray (this volume)

Figure 15. Relations of drainage stages to valley-fill stratigraphy and to interpreted drainage systems through time (after Bleuer, 1989, Table 3).

within the Mahomet sediments as well, a rough correlation of Lake Michigan Lobe [the Harmattan ice] and eastern lobe [Hillery-Tilton ice] events may be suggested.) That cap was locally removed (everywhere removed in Indiana) by late-glacial floods down shallow remnants of the Mahomet valley, or floods diverted into the Wabash via Attica. Such remnant Mahomet drainage persisted until final blockage by Illinoian drift associated with the Vandalia till (stage VIIb, Fig. 15). Appraisal of these possible physical relations depends on absolute knowledge of the geometries of the several well-known till units with respect to the main body of the Mahomet sand in Illinois.

Amid these same events, the Attica cutoff may have been

formed, and certainly it was filled. Many hypotheses are possible at this juncture, and most would be similar to Wayne's (1952) original ideas regarding diversion into the Wabash via the Danville valley. For instance, fan-head–like thickening of outwash associated with a stillstand of eastern ice (or by eastward-advancing Harmattan ice) could have caused diversion into the ancestral Wabash River via the Danville valley or the Attica cutoff. That position could be represented by the thickest and coarsest outwash pile in the Indiana part of the Mahomet, immediately west of the exit to the Attica cutoff (DNR 3a, Figs. 3D and 6A). The lowest 100 ft of fill of the cutoff at Attica (U.S. Highway 41 bridge borings) includes a till whose identity is not absolutely known. It resembles a basal facies of the Hillery till that is known in a nearby upland exposure. If the till is Hillery, the cutoff was open prior to, and possibly closed after, that ice advance (hypothesized stage Vb, Fig. 15). At this point, the model becomes tenuous and confused.

Once these valleys were filled to the present Wabash valley grade (~500 ft elevation), drainage down the Wabash bedrock valley was possible. The Vandalia caps the Mahomet sediments and records the end of any drainage along the Mahomet axis (stage VII, Fig. 15). The Vandalia advance may have caused that end and the penultimate creation of the Wabash River; or the Vandalia may simply fill minor western tributaries to an already existing Wabash.

During Sangamonian time, the ancestral Wabash flowed within the Battle Ground lowland at about the level of the present Wabash. This level, therefore, may have been controlled by the rock sill at Independence. The near coincidences of the Battle Ground lowland with: (1) positions of various ice margins of the Lake Michigan Lobe during the Wisconsinan; (2) the concentration of outwash derived from sources in the Tippecanoe and upper Wabash River basins; and (3) the location of a late-glacial and postglacial Wabash, surely are not coincidental. All these events are interrelated, and all in turn relate to a continuous thread of late Pleistocene drainage maintained within old valley routes that merged at Lafayette (stage VIII, Fig. 15).

The absolute chronology of the events noted above (Figs. 4, 15) is largely conjectural. The creation of the Wilshire–Blackford–West Lebanon association in the Marion valley plug is probably equivalent to Minford lacustrine sedimentation far up-valley, in terms of both position and reversed-polarity magnetic signature (Hoyer, 1976, 1983; Goldthwait, this volume; Bonnett and others, this volume). The event appears to have been separated from the subsequent creation of the Brookston-Mahomet-Harmattan association in the Metea-Mahomet inset by a significant erosional interval (drainage stage IVa, Fig. 15). Such an interval in Indiana must be equivalent, at least in part, to classic deep-stage incision in Ohio. If the West Lebanon till (and Minford deposits up-valley) is assumed to be greater than 0.7 m.y. old based on paleomagnetism, but less than 1.5 m.y. old based on amino-acid racemization values (Miller and others, 1987; W. D. McCoy, 1987, written communication), the West Lebanon event most likely equates with isotopic stage 22, the first and most clearly glacial event of the marine sequence within that time span (Shackleton and Opdyke, 1976, Fig. 1; Van Donk, 1976, Fig. 1). Through "counting-down" logic, type Sangamonian and type Illinoian events (the latter equating with the Vandalia till) commonly have been assumed to correspond with marine isotopic stages 5 and 6, respectively. Events relating to the Brookston, Harmattan, and Hillery tills fall somewhere within the above range.

REMAINING PROBLEMS

The stratigraphy of the fill of the so-called Teays, or the Lafayette Valley System in Indiana, does not appear to be particularly complex, but it is notable for the hiatuses that are represented within it. Interpreting what is missing at the unconformities within and atop the fill of the Lafayette system is impossible. In addition, it is possible that all the hiatuses have not yet been found. No cross-sectional profiles are available that might illustrate inset relations, a lack that is particularly significant in the broad expanse of the Mahomet section and in the broad, totally unexplored Frankfort lowland (Fig. 1).

It would be easiest to assume that any strata filling the system below the present grade of the Wabash River (given generally by the 525-ft elevation of the modern floodplain at the Lafayette hub) must be strictly fill of the old Marion-Mahomet valley, and that anything above that level probably relates to drainage into the lower Wabash Valley via the modern Independence rock sill. The discovery of the Attica cutoff, bypassing Independence at a lower elevation, suggests that the valley sediments *must* also record a drainage into the Wabash (such as that nebulously envisioned as stage V, Fig. 15). This distills to the simple question, "What, or how many different things, is the Mahomet Member?"

Various bothersome details remain, such as suggestions of scattered till-like deposits within the lowest Lake Marion sediments, the possible presence of paleosols (St. Marys, DNR 15) and the calcium-carbonate cementation common within the Blackford gravels, and the question of what, if not rock, confined the southeast end of Lake Anderson.

And finally, if a pre–West Lebanon glacial event produced the Marion valley, as suggested by other authors in this volume, where will its record be found and what will it look like? Might it look like the West Lebanon?

It would be remarkable if ancient valley history were as simple as is indicated by the stratigraphically based models presented here. The uneasy feeling remains that complicated interpretations may be as logical as simple ones. With as much as is now known, it is more imperative than ever before to appreciate what is not or cannot be known.

ACKNOWLEDGMENTS

Most of the original work reported here derives from a cooperative study by the Division of Water and the Division of Geological Survey (IGS), Indiana Department of Natural Resources (DNR), that culminated in a contractual Teays drilling program of 1977-1978. The acquisition of much new seismic information by the late Joseph Whaley (IGS), supervision of laboratory studies by Samuel Frushour (IGS), and the interchange with William Steen (DNR Division of Water) and Thomas Bruns (then DNR, Division of Water) throughout that program contributed greatly to this work.

The U.S. Geological Survey, Water Resources Division (USGS), particularly Daniel Gillies and Angel Martin, provided access to drilling sites associated with various projects in Howard and Cass Counties and in the upper White River basin. Jerry Hill and Marcia Taylor provided access to and information regarding the development of the Marion north well field. The cheerful cooperation of contracted drillers, Hobbs Drilling, Calvin Montgomery, then of Langfeldt and Payne Drilling (for USGS), and Rick Ortman of Ortman Drilling, Inc. (for USGS and DNR) is gratefully acknowledged. I am also indebted to numerous municipalities, other private drillers, firms, and individuals for access to many additional sites, not all of which are specifically discussed here. These include Darrel Leap for access to Purdue University tests, and Kevin Strunk, Wabash Valley Resources, for access to Palmer no. 1.

Several thoughts presented here (and in Bleuer, 1989) regarding nomenclature and valley history were developed following original presentations at the 1983 Geological Society of American national meeting and, particularly, following discussions with Henry Gray. I appreciate the thoughtful suggestions provided by reviewers Robert Shaver, Allan Schneider, Thomas Straw, and Richard C. Anderson.

REFERENCES CITED

Adams, J. M., Hinze, W. J., Brown, L. A., 1975, Improved application of geophysics to groundwater resource inventories in glaciated terrains: Purdue University Water Resources Center Technical Report 59, 62 p.

Arihood, L. D., 1982, Ground-water resources of the White River basin, Hamilton and Tipton Counties, Indiana: U.S. Geological Survey Water-Resources Investigations 82-48, 69 p.

Arihood, L. D., and Lapham, W. W., 1982, Ground-water resources of the White River basin, Delaware County, Indiana: U.S. Geological Survey Water-Resources Investigations 82-47, 69 p.

Ault, C. H., 1989, Map of Indiana showing directions of bedrock jointing: Indiana Geological Survey Miscellaneous Map 52.

Baker, V. R., Greeley, R., Komar, P. D., Swanson, D. A., and Waitt, R. B., Jr., 1987, Columbia and Snake River Plains, in Graf, W. L., ed., Geomorphic systems of North America: Boulder, Colorado, Geological Society of America, Centennial Special Volume 2, p. 403-468.

Belknap, D. F., and others, 1987, Later Quaternary sea-level chanes in Maine, in Nummendal, D., ed., Sea-level fluctuation and coastal evolution: Society of Economic Paleontologists and Mineralogists Special Publication 41, p. 71-85.

Blatchley, W. S., 1897, The petroleum industry in Indiana: Indiana Department of Geology and Natural Resources Annual Report 21, p. 27-96.

Bleuer, N. K., 1975, The Stone Creek section; A historical key to the glacial stratigraphy of west-central Indiana: Indiana Geological Survey Occasional Paper 11, 9 p.

—— , 1976, Remnant magnetism of Pleistocene sediments of Indiana: Proceedings of the Indiana Academy of Science, v. 85, p. 277-294.

—— , 1980, Correlation of pre-Wisconsinan tills of the Lake Michigan Lobe and Huron-Erie Lobe through the Teays Valley fill: Geological Society of America Abstracts with Programs, v. 12, p. 219.

—— , 1986, The nature of some glacial and other unconsolidated sedimentary sequences and their logging by natural gamma ray, in West, T. R., ed., Building on/with sedimentary bedrock: Proceedings 37th Annual Highway Geology Symposium, p. 209-219.

—— , 1989, Historical and geomorphic concents of the Lafayette Bedrock Valley System (so-called Teays Valley) in Indiana: Indiana Geological Survey Special Report 46, 11 p.

Bleuer, N. K., and Fraser, G. S., 1988, Mega-geometry of Wisconsinan glacial sequences [abs.]: Society of Economic Paleontologists and Mineralogists Midyear Meeting Abstracts, v. 5, p. 6.

—— , 1989, Lithofacies interpretations of glacial sediments of the Kankakee River Basin from downhole gamma-ray logs: Geological Society of America Abstracts with Programs, v. 21, p. 4.

Bleuer, N. K., and Melhorn, W. N., 1989, Glacial terrain models, north-central Indiana; The application of downhole logging to analysis of glacial vertical sequences, in Field Trip 2, Geological Society of America North-Central Section: p. 41-93.

Bleuer, N. K., and Moore, M. C., 1977, Basal till of so-called Kansan age in the upper Wabash Valley, Indiana: Geological Society of America Abstracts with Programs, v. 9, p. 576-577.

Bleuer, N. K., Melhorn, W. N., and Fraser, G. S., 1982, Geomorphology and glacial history of the Great Bend area of the Wabash Valley, Indiana, in 16th Annual Meeting, North-Central Section, Geological Society of America Field Guidebook: West Lafayette, Indiana, Purdue University, Department of Geosciences, 63 p.

Bleuer, N. K., Melhorn, W. N., and Pavey, R. R., 1983, Interlobate stratigraphy of the Wabash Valley, Indiana: 30th Annual Midwest Friends of the Pleistocene Field Conference Guidebook, 136 p.

Bownocker, J. A., 1899, A deep preglacial channel in western Ohio and eastern Indiana: American Geologist, v. 23, p. 178-182.

Burger, A. M., Keller, S. J., and Wayne, W. J., 1966, Map showing bedrock topography of northern Indiana: Indiana Geological Survey Miscellaneous Map 12.

Bruns, T. M., and Steen, W. J., 1980, Hydrogeology and ground-water availability in the preglacial Teays Valley, west-central Indiana [abstr]: Geological Society of America Abstracts with Programs, v. 12, p. 221.

—— , 1983, Hydrogeology of the Teays valley fill, north-central Indiana: Geological Society of America Abstracts with Programs, v. 15, p. 535.

Bruns, T. M., Logan, S. M., and Steen, W. J., 1985, Maps showing bedrock topography of the Teays Valley [eastern, central, and eastern parts], north-central Indiana: Indiana Geological Survey Miscellaneous Maps, 42, 43, 44.

Capps, S. R., 1910, The underground waters of north-central Indiana: U.S. Geological Survey Water-Supply Paper 254, 279 p.

Cummins, J. W., 1959, Probable surface of bedrock underlying the glaciated area in Ohio: Ohio Water Plan Inventory Report 10.

Dryer, C. R., 1920, The Maumee-Wabash waterway: Annals of the Association of American Geographers, v. 9, p. 41-51.

Fidlar, M. M., 1943, The preglacial Teays Valley in Indiana: Journal of Geology, v. 51, p. 411-418.

Fraser, G. S., and Bleuer, N. K., 1988, Internal architecture of the Valparaiso Moraine [abs.]: Society of Economic Paleontologists and Mineralogists Midyear Meeting Abstracts, v. 5, p. 20.

Frye, J. C., 1963, Problems interpreting the bedrock surface of Illinois: Illinois Academy of Science Transactions, v. 56, p. 3-11.

Gardner, M. C., 1979, Till stratigraphy of Tippecanoe County, Indiana [M.S. thesis]: West Lafayette, Indiana, Purdue University, 125 p.

Gillies, D. C., 1981, Ground-water potential of the glacial deposits near Logansport, Cass County, Indiana: U.S. Geological Survey Water-Resources Investigations 81-7, 98 p.

Gorby, S. S., 1886, Geology of Tippecanoe County: Indiana Department of Geology and Natural History Annual Report 15, p. 61–96.

Gray, H. H., 1982, Map of Indiana showing topography of the bedrock surface: Indiana Geological Survey Miscellaneous Map 35.

Horberg, L., 1945, A major buried valley in east-central Illinois and its regional relationships: Journal Geology, v. 53, p. 349–359.

——, 1950, Bedrock topography of Illinois: Illinois Geological Survey Bulletin 73, 111 p.

Hoyer, M. C., 1976, Quaternary valley fill of the abandoned Teays drainage system in southern Ohio [Ph.D. thesis]: Columbus, Ohio State University, 94 p.

——, 1983, Sediments of the Teays drainage system in southern Ohio: Geological Society of America Abstracts with Programs, v. 15, p. 600.

Indiana Geological Survey, 1970, Map of Indiana showing bedrock geology: Indiana Geological Survey Miscellaneous Map 16.

——, 1983, Map of Indiana showing topography on the bedrock surface: Indiana Geological Survey Miscellaneous Map 39.

Johnson, W. H., 1976, Quaternary stratigraphy in Illinois; Status and current problems, in Mahaney, W. C., ed., Quaternary stratigraphy of North America: Stroudsburg, Pennsylvania, Dowden, Hutchinson and Ross, Inc., p. 161–1976.

Johnson, W. H., Gross, D. L., and Moran, H. R., 1971, Till stratigraphy of the Danville region, east-central Illinois, in Goldthwait, R. P., Forsyth, J. L., Gross, D. L., and Pessl, F., Jr., eds., Till; A symposium: Columbus, Ohio State University Press, p. 184–216.

Johnson, W. H., Follmer, L. R., Gross, D. L., and Jacobs, A. M., 1972, Pleistocene stratigraphy of east-central Illinois: Illinois State Geological Survey Guidebook Series 9, 97 p.

Kayes, D. M., 1979, A gravity and seismic study of the buried Teays River, Benton, Tippecanoe, and Warren Counties, Indiana [M.S. thesis]: Bloomington, Indiana University, 170 p.

King, C. L., 1974, A study of buried glacial topography by the gravity method [M.S. thesis]: West Lafayette, Indiana, Purdue University, 83 p.

Lapham, W. W., 1981, Ground-water resources of the White River basin, Madison County, Indiana: U.S. Geological Survey Water-Resources Investigations 81-35, 112 p.

Lapham, W. W., and Arihood, L. D., 1984, Ground-water resources of the White River basin, Randolph County, Indiana: U.S. Geological Survey Water-Resources Investigations 83-4267.

Leverett, F., 1895, The preglacial valleys of the Mississippi and its tributaries: Journal of Geology, v. 3, p. 744–757.

Logan, W. N., 1920, Petroleum and natural gas in Indiana; A preliminary report: Indiana Department of Conservation Publication 8, 279 p.

Maarouf, A. M., and Melhorn, W. N., 1975, Hydrogeology of glacial deposits in Tippecanoe County, Indiana: Purdue University Water Resources Research Center Technical Report 61, 107 p.

Manos, C., 1961, Petrography of the Teays–Mahomet Valley deposits: Journal of Sedimentary Petrology, v. 31, p. 456–466.

McCabe, A. M., Dardis, G. F., and Hanvey, P. M., 1984, Sedimentology of a late Pleistocene submarine-moraine complex, County Down, Northern Ireland: Journal of Sedimentary Petrology, v. 54, p. 716–730.

McGinnis, L. D., 1968, Glacial crustal bending: Geological Society of America Bulletin, v. 79, p. 769–775.

McGinnis, L. D., and Heigold, P. C., 1973, Giant glacial grooves in the Meredosia Channel area, Illinois: Geological Society of America Abstracts with Programs, v. 5, p. 731–732.

——, 1974, A seismic refraction of the Meredosia Channel area of northwestern Illinois: Illinois State Geological Survey Circular 488, 19 p.

McGrain, P., 1950, Thickness of glacial drift in north-central Indiana: Indiana Flood Control and Water Resources Commission Circular 1.

Miller, B. B., McCoy, W. D., and Bleuer, N. K., 1987, Stratigraphic potential for amino acid ratios in Pleistocene terrestrial gastropods; An example from west-central Indiana, USA: Boreas, v. 16, p. 133–138.

Mitchell, J. P., 1984, Shallow seismic reflection data acquisition and processing techniques applied to delineation of buried bedrock topography [M.S. thesis]: West Lafayette, Indiana, Purdue University, 72 p.

Norris, S. E., and Spicer, H. C., 1958, Geological and geophysical study of the preglacial Teays Valley in west-central Ohio: U.S. Geological Survey Water-Supply Paper 1460-E, p. 199–232.

Norris, S. E., and Spieker, A. M., 1966, Ground-water resources of the Dayton area, Ohio: U.S. Geological Survey Water Supply Paper 1808, 167 p.

Phinney, A. J., 1890, The natural gas field of Indiana: U.S. Geological Survey Annual Report 11, pt. 1, p. 579–742.

Powell, R. D., 1983, A model for sedimentation by tidewater glaciers, in Gold, L. W., ed., Proceedings of the Symposium on Processes of Glacier Erosion and Sedimentation: Annals of Glaciology, v. 2, p. 129–134.

Rosenshein, J. S., 1958, Ground-water resources of Tippecanoe County, Indiana: Indiana Division of Water Resources Bulletin 8, 38 p.

Shackleton, N. J., and Opdyke, N. D., 1976, Oxygen-isotope and paleomagnetic stratigraphy of Pacific core V28-239 late Pliocene to latest Pleistocene, in Cline, R. M., and Hays, J. D., eds., Investigation of Late Quaternary paleoceanography: Geological Society of America Memoir 145, p. 449–464.

Shaw, J., 1989, Drumlins, subglacial meltwater floods, and ocean responses: Geology, v. 17, p. 853–856.

Smith, B. S., Hardy, M. A., and Compton, E. J., 1985, Water resources of Wildcat Creek and Deer Creek basins, Howard and parts of adjacent counties, Indiana, 1979–82: U.S. Geological Survey Water-Resources Investigative Report 85-4076, 92 p.

Stephenson, D. A., 1967, Hydrogeology of glacial deposits of the Mahomet Bedrock Valley in east-central Illinois: Illinois State Geological Survey Circular 409, 51 p.

Stout, W., Ver Steeg, K., and Lamb, G. F., 1943, Geology of water in Ohio: Ohio Geological Survey, 4th series, Bulletin 44, 694 p.

Thornbury, W. D., 1948, Preglacial and interglacial drainage of Indiana [abs.]: Geological Society of America Bulletin, v. 59, p. 1359.

Thornbury, W. D., and Deane, H. L., 1955, The geology of Miami County, Indiana: Indiana Geological Survey Bulletin 8, 49 p.

Tight, W. G., 1903, Drainage modifications in southeastern Ohio and adjacent parts of West Virginia and Kentucky: U.S. Geological Survey Professional Paper 13, 111 p.

Van Donk, J., 1976, O^{18} record of the Atlantic Ocean for the entire Pleistocene Epoch, in Cline, R. M., and Hays, J. D., eds., Investigation of Late Quaternary paleoceanography: Geological Society of America Memoir 145, p. 147–163.

Ver Steeg, K., 1946, The Teays River: Ohio Journal of Science, v. 46, p. 297–307.

Waitt, R. B., Jr., 1985, Case for periodic, colossal, jokulhlaups from Pleistocene glacial Lake Missoula: Geological Society of America Bulletin, v. 96, p. 1271–1286.

Walker, R. G., 1984, Turbidites and associated coarse clastic deposits, in Walker, R. G., Facies models, 2nd. ed.: Geoscience Canada Reprint Series 1, p 170–188.

Wayne, W. J., 1952, Pleistocene evolution of the Ohio and Wabash Valleys: Journal of Geology, v. 60, p. 575–585.

——, 1956, Thickness of drift and bedrock physiography of Indiana north of the Wisconsin glacial boundary: Indiana Geological Survey Report of Progress 7, 70 p.

——, 1963, Pleistocene formations in Indiana: Indiana Geological Survey Bulletin 25, 85 p.

Wayne, W. J., and Thornbury, W. D., 1951, Glacial geology of Wabash County, Indiana: Indiana Geological Survey Bulletin 5, 39 p.

MANUSCRIPT ACCEPTED BY THE SOCIETY JUNE 29, 1990

Printed in U.S.A.

Aquifer systems of the buried Marion-Mahomet trunk valley (Lafayette Bedrock Valley System) of Indiana

N. K. Bleuer
Geological Survey Division, Indiana Department of Natural Resources, Bloomington, Indiana 47405
W. N. Melhorn
Department of Earth and Atmospheric Sciences, Purdue University, West Lafayette, Indiana 47907
W. J. Steen and T. M. Bruns*
Division of Water, Indiana Department of Natural Resources, Indianapolis, Indiana 46241

ABSTRACT

Groundwater resources associated with sediments filling the Marion and Mahomet Valley Sections of the Lafayette Bedrock Valley System vary from miniscule to substantial, reflecting the wide range of glacigenic aquifer facies contained in the fill. These aquifer facies include braid-stream deposits that range from thin units within till sequences to immense, valley-filling masses. Also included is a variety of proximal to distal, subaerial to subaqueous, fan and fan-delta deposits; these range from thick masses of ice-proximal, cobbly rubble interspersed with thin diamicts and clays, to thin, discontinuous lentils of sand confined within lacustrine clays.

Valley-fill aquifers are confined by capping till units, except where exhumed at the crossings of the Maumee-Wabash Trough (modern Wabash River Valley). A variety of aquifers typically are available within the valley-capping sediments; for this reason, much of the deep valley-fill has not been extensively explored or developed. Some valley-fill aquifers are so thin and/or deeply buried that their exploitation is unlikely, but others are so thick and areally extensive that exploitation easily can support sustainable yields of tens of millions of gallons per day.

BACKGROUND

We summarize here the character of the water resources of the various parts of the Marion and Mahomet Valley Sections of the Lafayette Bedrock Valley System in Indiana. The data derive primarily from the exploration program of the Indiana Department of Natural Resources (DNR), Division of Water and Division of Geological Survey (described in Bleuer, this volume), and from other studies partly derived from that program (Bruns and Steen, 1980, 1983). Herein, discussion is organized in the context of the sedimentary facies and rock stratigraphy as described by Bleuer in this volume. (For an index to designated valley sections and counties, see Bleuer, Fig. 1, this volume, and Bleuer, 1989, Plate 1).

Prior to the DNR program, the character of groundwater resources associated with the Marion-Mahomet trunk valley could only be estimated. Ideas were pervasive of immense aquifer systems, as were assumed to be associated with buried valley systems in general. However, proof of the existence of such a system in Illinois (Stephenson, 1967) was countered by proof of absence of such a system in Ohio (Norris and Spicer, 1958). Knowledge of groundwater resources in Indiana was restricted to local areas where valley-fill sediments have been partly exhumed, such as beneath the Maumee-Wabash Trough (modern Wabash River Valley) at Lafayette, Tippecanoe County, and at Peru, Miami County. Aspects of the valley at Lafayette are discussed by Rosenshein (1958), Maarouf and Melhorn (1975), Pohlmann (1987), Weinreb (1987), and Harvey (1990). Elsewhere, intensive exploration has delineated resources of the Logansport area (U.S. Geological Survey—Cass County data set used in Bleuer, this volume; Gillies, 1981; also summarized in Noel, 1978), but

---

*Present address: Indianapolis Water Company, P.O. Box 1220, Indianapolis, Indiana 46206.

Bleuer, N. K., Melhorn, W. N., Steen, W. J., and Bruns, T. M., 1991, Aquifer systems of the buried Marion-Mahomet trunk valley (Lafayette Bedrock Valley System) of Indiana, *in* Melhorn, W. N., and Kempton, J. P., eds., Geology and hydrogeology of the Teays-Mahomet Bedrock Valley System: Boulder, Colorado, Geological Society of America Special Paper 258.

knowledge of other areas was limited to a variety of older, regional studies (Capps, 1910; Wayne and Thornbury, 1951; Thornbury and Deane, 1955; Wayne, 1956). Then, as now, available data consisted of records of water wells that generally terminated above true valley fill, and records of oil wells from the area of the "deep drive" of the old Trenton Field, in Grant, Blackford, Jay, and Adams Counties, which recorded virtually no unconsolidated stratigraphy.

Following the DNR program, municipal wells or well fields and other projects have provided formation characteristics in various parts of the valley; these include sites in or near Templeton, Oxford, Otterbein, and Lafayette (Mahomet Section), Peru, (Peru Segment), Marion (Geneva Segment), and Berne (St. Marys Bend Segment).

AQUIFER SYSTEMS OF THE VALLEY FILL

General setting

The availability of groundwater within fill of the Marion-Mahomet trunk valley varies from miniscule to substantial, depending on characteristics of both the aquifers and surrounding materials. Bedrock beneath and aside the valley varies from Ordovician shale on the east, through Lower and Middle Paleozoic carbonate rocks centrally, to Mississippian siltstone and shale on the west (Bleuer, this volume, Fig. 2). Confining materials above the valley, through which most recharge occurs, range from varying thicknesses of clayey till of the Lagro Formation on the east, to stacked, loamy tills of several ages on the west. Where unconfined, as in the area of juxtaposition with the Wabash River Valley of central Indiana (the Maumee-Wabash Trough), the aquifers lie in zones of regional groundwater discharge to the Wabash River; concomitantly, the upper aquifers are susceptible to direct, rapid, induced recharge from the surface.

Aquifers within the valley-fill sediments correspond to the coarser-grained facies of various fluvial outwash and subaqueous fan environments (Bleuer, this volume, Figs. 6, 7, and 9). The aggregate thickness of coarse-grained valley fill (Fig. 1) exaggerates bona fide aquifer thickness in many locales, but nonetheless grossly outlines the highly variable distribution of aquifer environments.

Regional aquifers capping the valley

Various intra-till and sub-till, sand-and-gravel aquifers are associated with drift sequences overlying true valley-fill sediments. One such aquifer, not yet adequately delineated, is an intra- or sub-Wisconsinan aquifer associated with the "loblolly" of eastern Indiana (Bleuer, 1989, Plate 2), a problem discussed below in more detail. The most significant of this type of aquifer is associated with the La Fontaine Segment of the Marion Valley section in Wabash, Grant, and Blackford Counties, where a thick, gravelly sand aquifer is widespread. This unit lies above true valley fill; the base apparently corresponds to the level of the ⩽700-ft, flanking bedrock surface (primarily area D, Fig. 1; DNR 10-18 in Fig. 6C, D of Bleuer, this volume). As this unit lies above "true" valley fill, as defined by the ∼700-ft lip of the inner valley trench, and by association with red, fine-grained sediments, the unit's distribution is not included in sand thickness as shown in Figure 1. This aquifer unit is interpreted as a stack of braid stream and/or fan deposits, possibly of Wisconsinan age (Bleuer, this volume). The unit underlies thin, loam-textured till, probably equivalent to the basal till of the Wisconsinan Trafalgar Formation and thicker, clayey till of the Lagro Formation, and overlies a variety of reddish sediments of the Blackford Member.

The unit described is a major regional aquifer, although data are insufficiently widespread to predict lateral relations or internal characteristics. Confined regionally, the aquifer appears to be interconnected locally with discontinuous aquifers within underlying valley fill.

In a specific example, within the city of Marion's recently developed north well field, located atop the outer crest of the Mississinewa Moraine near the Grant-Wabash County line, the valley-capping aquifer (unit A, Fig. 2B) lies beneath an aquitard consisting of basal and supraglacial clayey tills of the Lagro Formation and thin, loamy Trafalgar(?) till. This aquifer in turn overlies a variety of thinner, intralacustrine aquifers nested within the true valley-fill sequence (unit[s] B, Fig. 2b), described below. The capping aquifer is, for the most part, the uppermost of six aquifer units identified and mapped within the field by project consultants (Stremmel and Hill, 1985).

The field itself consists of a north and south field, two clusters of three production wells each, ranging in depth from 160 to 230 ft. The southern cluster taps only the upper, valley-capping aquifer, which is more than 80 feet thick (e.g., MTW 21, Fig. 2b). Thin, silty, but progressively coarsening-upward sediments encountered in several wells penetrating this unit are presumed to represent local, deltaic fill of braid stream plugs—these clearly are not widespread aquitards. The northern well cluster taps both the upper aquifer and a lower system of aquifers, which consist of generally thinner granular units that are highly variable in character and position (e.g., MTW 20, 21, Fig. 2b). These lower units are associated with a local, coarse-grained deltaic facies of the Blackford Member, as discussed subsequently. The upper aquifer and a basal aquifer were also encountered in the old Marion northeast field, located in a tributary valley south of the Marion-Mahomet trunk valley.

The upper aquifer and various lower aquifers in Marion's north well field are undoubtedly in direct or indirect hydraulic contact and are interactive, as discussed below. Available pumping data suggest that local transmissivities of the upper aquifer may be as great as 128,000 g/d/ft (gallons/day/foot), or 1.8×10^{-2} m^2/s (meters^2/second), with a regional average of 85,000 g/d/ft (1.2×10^{-2} m^2/s) (Stremmel and Hill, 1985). Drawdown in production wells developed in all aquifers presently has stabilized, with a total field production of about 2.5 Mg/d (Roger Pettijohn, oral communication).

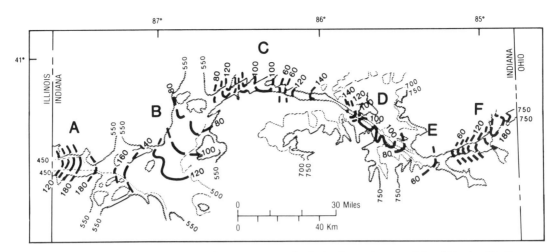

Figure 1. Generalized, aggregate-thickness contours (in feet) of granular materials within the fill of the Marion-Mahomet trunk valley. Sediments include silty sand through coarse gravel; therefore contours best represent geometries of the sedimentary environments, not necessarily a bona fide thickness of aquifer materials. Convex, increasing contours are drawn to suggest positive, fan-like geometries emanating from source areas (central and eastern sections); concave, increasing contours are drawn to suggest the possibility of negative, erosional geometries (western sections) (see Bleuer, this volume, Figs. 6 and 7).

Regional aquifers filling the valley

The Mahomet Member makes up most of the volume of the Mahomet Valley Section, west of Lafayette in Tippecanoe, Benton, and Warren Counties (area A, Fig. 1; also DNR 1-5, Figs. 6A and 7 of Bleuer, this volume). The unit is dominantly either sand and fine gravel or cemented, coarse-grained gravel, depending on location; data are inadequate to generalize characteristics for the entire 2 to 6 mi width of the valley. Thin, discontinuous tills are interspersed within gravel in the Lafayette area, but westward the member appears to be uninterrupted sand and gravel, lacking regional aquitards or even local discontinuities such as those noted within the Marion capping aquifer. Thus the unit is interpreted as stacked braid-plain deposits. These true valley-fill materials are overlain directly by valley-train sand and gravel (and included Wisconsinan tills), which make up the terraces that flank the modern Wabash River Valley, within the Maumee-Wabash Trough; elsewhere, the valley fill is capped by a variety of Illinoian and younger till sequences (Bleuer, this volume, Fig. 6A).

The Mahomet Member is a major regional aquifer. It is confined, except where exhumed at the crossing of the Maumee-Wabash Trough. In that area, the aquifer is under water-table conditions, and gradients reflect discharge into the modern valley; upper, till-capped parts of this aquifer lying beneath adjacent uplands are commonly dry, reflecting localized water-table conditions in those areas as well.

Available pumping data suggest that transmissivities of the aquifer west of Lafayette range from 100,000 to 300,000 g/d/ft (1.5×10^{-2} to 4.4×10^{-2} m^2/s) (Bruns and Steen, 1983). In the specific example of the small community of Oxford, in southern Benton County west of Lafayette, the primary public water supply is derived from a single well developed within the thick, gravel aquifer. The community experienced periodic water supply problems for years, before drilling a well about 4 mi south of town near the north edge of the Mahomet valley, at a site selected on the basis of the DNR study. At this site, the valley is filled with slightly more than 100 ft of sand and gravel (Fig. 3c). The well, which yields more than 0.72 Mg/d with a drawdown of 7 ft, will provide Oxford's water needs for years to come.

A similar situation was noted at Otterbein, where the town previously had tapped a shallow sand-and-gravel aquifer (50 to 60 ft deep), yielding very hard, iron-rich water. A new production well developed at existing water facilities at the town's center, located atop the northernmost edge of the Mahomet valley fill, penetrates about 100 ft of the sand and gravel beneath the shallower aquifer (Fig. 3b). This lower aquifer yields more than 600 gpm with less than 5 ft of drawdown; iron content of the water is slightly more than 1 mg/l.

The Greater Lafayette community is the major groundwater consumer tapping the western part of the trunk-valley aquifer. Principal withdrawal is from municipal wells in Lafayette and West Lafayette, a Ranney well system 3 mi southwest at the Eli Lilly pharmaceutical plant, and the Purdue University well field in West Lafayette.

Lafayette's main well field, near the east bank of the Wabash River and receiving some recharge therefrom, contains 10 wells, whose maximum spacing between any wells is only about 230 ft. The wells are 90 to 100 ft deep and penetrate an average, screened aquifer thickness of 20 to 25 ft. Three recently

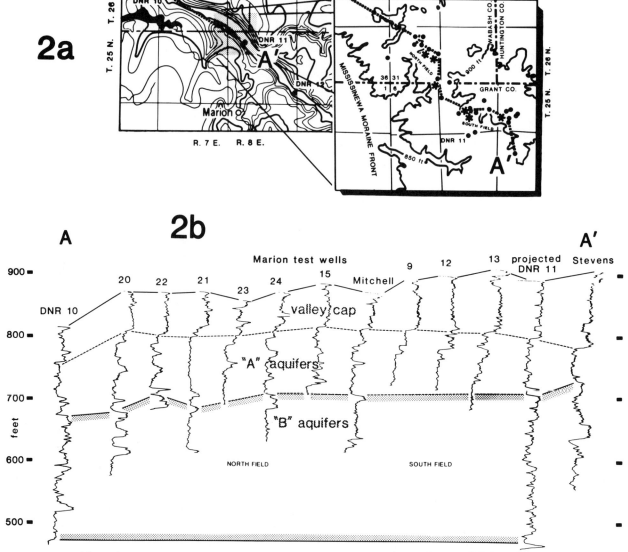

Figure 2. Anatomy of aquifers of the Marion north well field, La Fontaine Segment, Marion Valley Section. a, Bedrock topography and location of the lines of cross section, Grant and Wabash Counties. Bedrock contours at 50-ft intervals (small scale, from Gray, 1982) define general position of the trunk valley. Surface contours (large scale, modified from USGS La Fontaine 7½-min quadrangle) define the front (<850 ft) and crest (>900 ft) of the Mississinewa Moraine. Production wells denoted by asterisk. b, Abbreviated cross section A-A', interpreted. Generalized sedimentary environments of aquifers of the north field area (MTW 20-22) related to background valley fill (DNR 10-11). c, Cross section B-B'. Gamma-ray logs from selected test holes within the wellfield, showing the distribution of aquifer materials in units A and B. Other bounding materials include variable Lagro Formation ablation deposits (La), massive basal till of the Lagro Formation (Lb), and basal till and proglacial sequences of Trafalgar Formation (T). "B" aquifers are embedded within finer-grained sediments of the Blackford Member, Banner Formation (Bleuer, this volume). (Note for Fig. 2b and c: hole spacings are not to scale). Marion wellfield data are derived from studies by Stremmel and Hill and Ortman Drilling for the city of Marion. MTW test holes were logged through 2-in PVC casing; Stevens and Mitchell residential wells through 5-in PVC casing; DNR holes through 4-in steel drill pipe; sharp inflections at 20-ft. intervals reflect pipe joints.

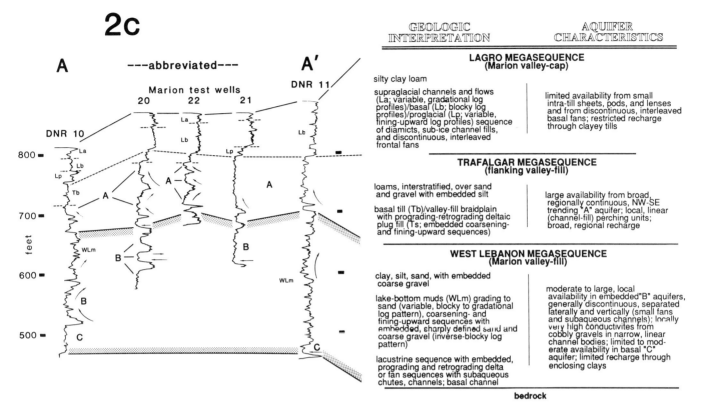

developed wells, spaced about 1,000 ft apart, are about 4,000 ft farther north and penetrate a similar thickness of aquifer. Seven West Lafayette wells occupy essentially an opposing position on the west bank of the Wabash River and draw from the same valley-fill aquifer.

The saturated thickness of the semi-confined aquifer averages about 100 ft, but ranges from less than 50 ft to more than 140 ft within the general Lafayette lowland. Several measurements or calculated estimates are available for hydraulic conductivity and transmissivity. Hydraulic conductivity ranges from 1,197 to 1,795 g/d/ft^2 (5.6×10^{-4} to 8.5×10^{-4} m/s) for Lafayette municipal wells, and a conservatively rated 2,244 g/d/ft^2 (1.1×10^{-3} m/s) for the Purdue field (Pohlmann, 1987). Corresponding transmissivities for the two fields, reflecting varying positions within the aquifer, range from 97,240 to 279,750 g/d/ft (1.4×10^{-2} to 4.0×10^{-2} m^2/s). Transmissivities associated with the vast, undeveloped reserves of the Mahomet trunk valley, several miles northwest of West Lafayette, are estimated as exceeding 350,000 g/d/ft (5.1×10^{-2} m^2/s).

The aquifer was severely tested during a rare event associated with the 1988 drought. Culminating on July 14, all 13 Lafayette wells had been on line for 55 hr. Average pumpage for the period was 19.8 Mg/d with drawdowns of 7 to 26 ft—total rated pumping capacity for the wells is 21 Mg/d. During the same interval, West Lafayette, Eli Lilly, and Purdue University pumped 6.9, 10, and 6.8 Mg/d, respectively, with no evidence of exceeding safe yield. Clearly, the valley fill at Lafayette is a major aquifer.

Regional intratill aquifers

The Brookston Till Member of the Jessup Formation (Bleuer, this volume) appears to be broken internally by at least one intratill sand and gravel body within the Battle Ground Lowland Section, Tippecanoe and Carroll Counties. This body expands and coalesces with the basal Brookston aquifer westward to become part of the valley-filling Mahomet aquifer (area B, Fig. 1; also DNR 5-8, Figs. 6A and 7 of Bleuer, this volume). If interpreted correctly as a proximal-distal progression of fan and braided outwash plain, the aquifer probably splits eastward into multiple units, though maintaining uniform thickness through the central Battle Ground lowland.

Locally where coarsest, and regionally where continuous, the intratill Brookston unit is a significant aquifer. However, the valley fill has not been extensively explored because Delphi, the largest town in the area, obtains adequate water from shallow limestone wells on the valley's southeast edge. The intra-Brookston unit is generally confined, but may be unconfined where exhumed beneath the Maumee-Wabash Trough, which traverses the long axis of the Battle Ground lowland. With the Tippecanoe and Wabash Rivers as control points for groundwater discharge, water levels in aquifers of the Brookston area are deeper than expected in most till-capped areas of central Indiana.

A prominent sand and gravel aquifer, below the confining surficial till, is believed to extend northeastward from the Brookston area to west of Delphi, and northward to Lake Cicott

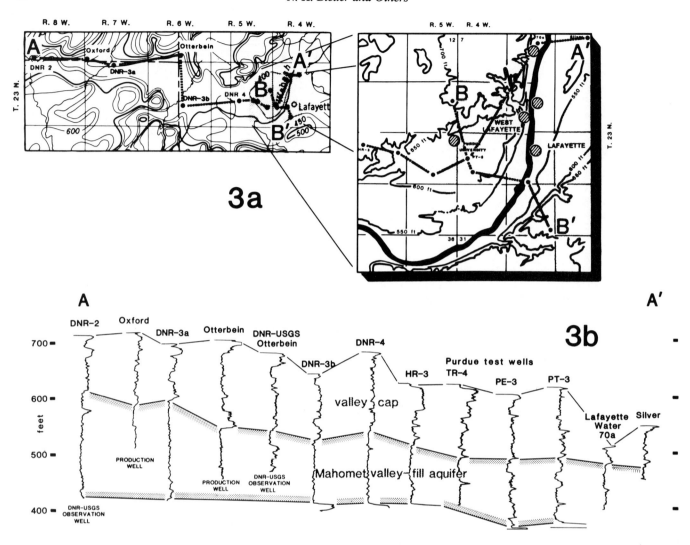

Figure 3. Anatomy of the massive valley-fill aquifer, Mahomet Valley Section. a, Location of lines of cross section, Benton and Tippecanoe Counties. Bedrock contours at 50-ft intervals (small scale, from Gray, 1982) define general position of the trunk valley. Surface contours (large scale, modified from USGS Lafayette West 7½-min quadrangle) define the Maumee-Wabash Trough (<650 ft), inset Wisconsinan terraces (650 to 550 ft), Wabash River Valley floodplain (<500 ft), and surrounding uplands. Well fields shown by hachured circles. b, Cross section B-B′, interpreted. Sedimentary environments of aquifers of the valley fill and cap of the Lafayette area, Mahomet Valley Section. Aquifer A is the massive, coalesced valley-filling aquifer of western Indiana and eastern Illinois (Mahomet Member in Indiana, Mahomet Sand in Illinois). c, Cross section A-A′. Gamma-ray logs from selected test holes within the valley, showing the broader distribution of aquifer materials in units A and B. Other bounding materials include a variable Wisconsinan drift cap (Bleuer, this volume, Figs. 6 and 7). (Note for Figs. 3b and c: hole spacings are not to scale). Site at Oxford is the producing well; sites at Otterbein and Lafayette (70a) are test wells near producing wells. DNR 3 and 4 are derived from the DNR study (Bleuer, this volume). Purdue tests (PE and PT) are derived from Purdue University thesis/contract studies. State Road 63 bridge and the Silver well are derived from a construction test and a domestic well, respectively. Oxford and Otterbein holes were logged through 12-in steel and 4-in PVC casing, respectively; Purdue tests through 2-in PVC; DNR holes and SR 26 bridge hole through 4-in steel drill rod; sharp inflections at 20-ft intervals reflect rod joints.

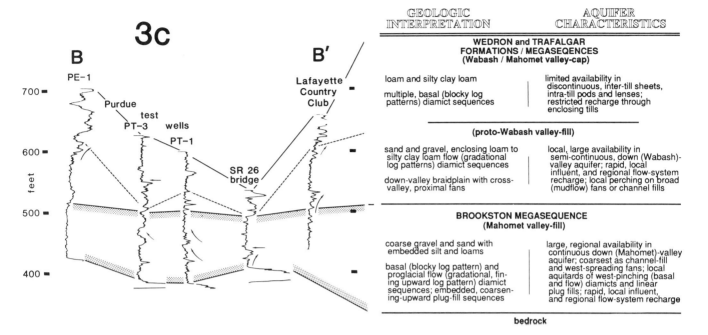

(near DNR 8, Fig. 6A of Bleuer, this volume). This aquifer, which in part may be an intra- or basal-Wisconsinan unit, is utilized by several irrigation wells southwest of Lake Cicott, with individual yields exceeding 1 Mg/d.

Basal aquifers

A basal sand and gravel, which is locally very coarse, cobbly, and cemented, underlies the Brookston till within the valley fill of the Battle Ground Lowland Section and the Logansport Bend Section, in Tippecanoe, Carroll, and Cass Counties. (This unit is a primary component of aggregate thickness contours, area B, Fig. 1; see also DNR 5-USGS 93, Figs. 6A, B and 7 of Bleuer, this volume.) The unit is locally very coarse grained. Cemented sand and gravel, containing boulders and local limestone (colluvial?) rubble is particularly common west of Delphi, where a thickness of 10 to 30 ft may be expected. If interpreted correctly as a pro-Brookston fan and through-flowing, valley-filling outwash train, the unit may be continuous down-valley.

This unit may encompass a locally significant aquifer. It is confined except where it merges with the unconfined part of the Mahomet aquifer near Lafayette (area B, Fig. 1). Although no specific information is available on aquifer capability, it must vary greatly depending on cementation and sediment characteristics. Yields of 0.07 to 0.6 Mg/d probably can be expected for appropriately constructed wells.

In contrast to the Brookston-Delphi area, in the eastern segments of the Marion valley a basal, fine-grained sand and gravel lies at the base of the Blackford Member, the assemblage of glaciofluvial and glaciolacustrine sediments that there fills the valley. (This unit is a contributor to aggregate thickness contours, areas C, D, E, F, Fig. 1; see also DNR 10-15, Figs. 6D and 7 of Bleuer, this volume.) In contrast to discontinuous units within the fluvio-lacustrine fill, this basal sand and gravel appears ubiquitous; as such, it has been interpreted as partly fan and partly through-flowing, braided outwash predating the damming of the valley.

If interpreted correctly, the unit probably is narrow, but maintains longitudinal continuity. Nevertheless, the unit must be considered a minor, localized aquifer. In fact, of all aquifers described herein, the basal Blackford aquifer is probably the least known or utilized. A common rule of thumb among drillers in central Indiana is that significant aquifers are not found once the "red clay" has been encountered. This is probably so, because the West Lebanon ice commonly rode up local valleys, such that reddish till and/or lacustrine clays are the dominant sediment capping limestone bedrock. Because of this experience, the great depth, and the fact that the clays are hard and slick, making rotary drilling very slow, this potential aquifer is not likely to be explored extensively. The unit is tightly confined by the capping lacustrine clays, although probably the hydrology merges with that of the surrounding limestone bedrock.

Complex, ice-proximal fan aquifers filling the valley

The western, ice-proximal facies of the Blackford Member (Bleuer, this volume), which fills the Peru Segment of the Marion Valley Section in Cass and Miami Counties, is a complex of interstratified, coarse-grained to bouldery outwash associated with a variety of red tills and clays. The aquifers originated as subaerial and subaqueous fan sediments, whose aggregate thickness reflects their introduction into the trunk valley from discrete sources to the northwest, north, and northeast. (The unit is a primary component of the up-valley and down-valley lobate, aggregate thickness contours, area C, Fig. 1; see also USGS 92-DNR 10, Figs. 6C, D and 7 of Bleuer, this volume.)

The complex distribution and characteristics reflect the several sources of materials; thickest and variably coarse-grained in the Logansport area on the west, they laterally merge with intra- and sub-Brookston aquifers. But overall these sediments are coarsest (cobbly/bouldery) at the crossing of the Maumee-Wabash Trough (the modern Wabash River Valley) in the Peru area on the east (Fig. 6C of Bleuer, this volume; Bleuer, 1989, Plate 1). At this crossing, coarse-grained, cobbly deposits related to the late Wisconsinan flood discharges are superposed atop the exhumed valley-fill aquifer (Fraser and Bleuer, 1988), although the depth of that flood scour is not known, and discrimination of Wisconsinan and pre-Wisconsinan sediments is not possible. Eastward, the units descend, thin abruptly, and then pinch out. Thus if interpreted correctly, the aquifer is of regional extent and thick, in aggregate; however, individual elements of the western facies must be assumed to be discontinuous and of highly variable character and thickness.

Aquifers of the western areas are confined, and their water-bearing capabilities are great. To the east, the aquifer is unconfined where it is exhumed and in continuity with coarse-grained sediments of the Maumee-Wabash Trough, between Peru and Richvalley* in eastern Wabash County, and capabilities are greatest; available pumping data in that area suggest that transmissivities range from 250,000 to 300,000 g/d/ft (3.6×10^{-2} to 4.3×10^{-2} m^2/s) (Bruns and Steen, 1983).

A specific example of an as-yet untapped resource is aquifers above and within the valley fill in the vicinity of Logansport, Cass County (USGS–Cass County data set of Bleuer, this volume, Figs. 2 and 6B). Logansport presently pumps about 4 Mg/d directly from the Eel River for municipal use. However, the uplands generally north of the community are consistently underlain by an aquifer complex that has been modeled as a semi-confined, three-tier aquifer system (Gillies, 1981). Two upper aquifers each are less than 20 ft thick, discontinuous, and interlayered with till; these are interpreted as intra-Wisconsinan outwash bodies by Bleuer (this volume). These aquifers crop out along valley walls and therefore are suitable only for limited domestic withdrawals. The basal aquifer, associated with true valley fill (the western, ice-proximal facies of the Blackford Member of Bleuer, this volume), averages 80 ft thick along the valley axis, but rarely is reached by domestic wells; the north-south valley width is only about two miles. Favorable sites for future Logansport production lie about 4 mi north of the city. Gillies' models have simulated a withdrawal of 10 Mg/d in areas of high transmissivity (calculated average transmissivity is 119,680 g/d/ft or 1.7×10^{-2} m^2/s, based on 34 test wells). Simulation indicated no constraints on such withdrawal, but also suggested drawdowns as great as 35 ft in the deep aquifer, and simultaneous drawdowns of as much as 20 ft in the commonly used shallow aquifers, separated from the former by a till aquitard.

In the area between Peru and Richvalley, deposits of thick, bouldery, cemented sand and gravel lie directly beneath the Maumee-Wabash Trough, and overlie thick sand and gravel. Aquifer thicknesses as great as 200 ft are recorded, and the gravel is interrupted only locally by thin, sandy clays that probably represent discontinuous mudflow bodies related to the West Lebanon till (Fig. 2c; also Figs. 6B and 11 of Bleuer, this volume). This aquifer is utilized by the city of Peru, whose newest production wells, located within the city limits on the south edge of the valley, are capable of producing 2.9 Mg/d each.

In a related example, the main well field supplying the city of Wabash is located about 3 mi south of that city in a short, buried tributary valley. This valley is one of several east of Richvalley, where the Marion-Mahomet trunk valley departs from the Maumee-Wabash Trough. These local, bedrock tributaries, entering the main valley from the north, are partly filled with coarse-grained colluvium, sand, and gravel. Wabash wells yield in excess of 1,000 gpm. The original Wabash city well field was located in a somewhat longer, better-defined bedrock tributary extending from the city southwestward to the main valley. Wells in this field, which originally flowed, yielded more modest quantities of water.

Deposits representing a similar, but less extreme, ice-proximal fan environment occur in the St. Marys Bend and upper Geneva Segments of the Marion Valley Section, in Adams County; this eastern ice-proximal facies thins and pinches abruptly westward (area F, Fig. 1; also DNR 14-St. Marys, Figs. 6D and 7 of Bleuer, this volume). Coarse-grained, cobbly sediments are recorded only in a single test hole (DNR 15), but may be more extensive eastward in Ohio. The coarse-grained sediments apparently grade westward into finer-grained deltaic(?) sands embedded within a lacustrine sequence, similar to those of the Marion area as discussed below.

The uppermost aquifers of this area may be, in part, of Wisconsinan age. The Geneva Segment is nearly paralleled at the modern surface by the unusual landscape of the "loblolly" (Fig. 4a; Bleuer, 1989, Plate 2). This northeast-southwest–lineated, medium-relief landscape is suggestive of a complex of collapsed tunnel valleys; thus, the uppermost aquifers could relate to subglacial discharge that preferentially cut into valley fill rather than adjacent limestone bedrock. However, these aquifers, because of their characteristics and positions, are difficult to distinguish from the uppermost valley-fill aquifer.

In specific examples, the city of Decatur's well field east of Berne, and the new Berne well field to the southwest (Fig. 4b) encountered coarse-grained, cobbly sand and gravel from about 110 to 150 ft in depth. These materials probably are related to a similar coarse-grained unit encountered in DNR 15 farther east (Fig. 4a). Production wells in these fields yielded from 1.4 to 3.0 Mg/d with drawdowns as great as 40 ft. This confined aquifer, when pumped by the city of Decatur, and later Berne, resulted in a decline in water level that affected a number of

---
*The abrupt broadening of the Maumee-Wabash Trough at Richvalley is the classic example of modern drainage crossing a buried rock valley (Thornbury, 1969, Fig. 2.8). This area is the easternmost surface exposure, and expression of, Marion-Mahomet valley-fill sediments and aquifers.

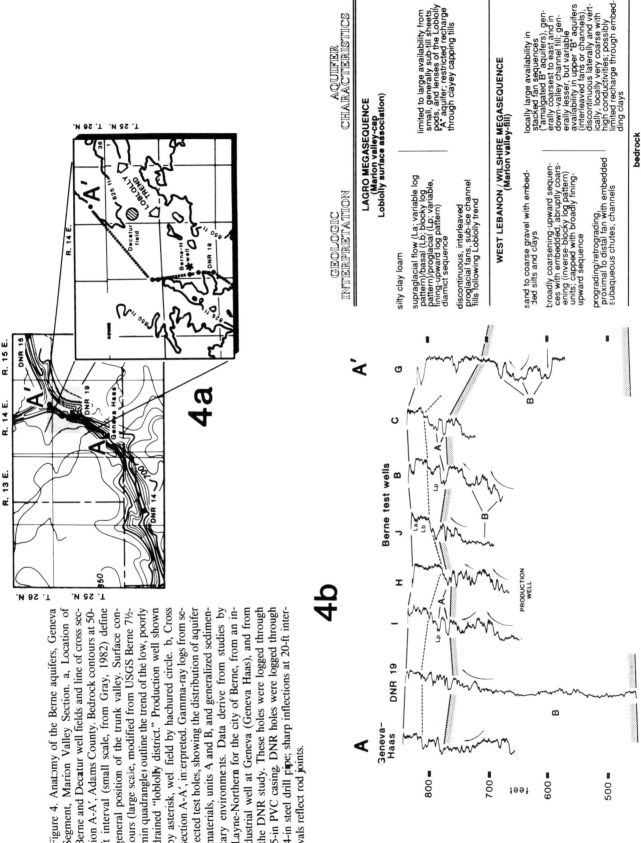

Figure 4. Anatomy of the Berne aquifers, Geneva Segment, Marion Valley Section. a, Location of Berne and Decatur well fields and line of cross section A-A', Adams County. Bedrock contours at 50-ft interval (small scale, from Gray, 1982) define general position of the trunk valley. Surface contours (large scale, modified from USGS Berne 7½-min quadrangle) outline the trend of the low, poorly drained "loblolly district." Production well shown by asterisk, well field by hachured circle. b, Cross section A-A', interpreted. Gamma-ray logs from selected test holes, showing the distribution of aquifer materials, units A and B, and generalized sedimentary environments. Data derive from studies by Layne-Northern for the city of Berne, from an industrial well at Geneva (Geneva Haas), and from the DNR study. These holes were logged through 5-in PVC casing. DNR holes were logged through 4-in steel drill pipe; sharp inflections at 20-ft intervals reflect rod joints.

nearby household wells. Deepening of pump settings, and in some cases the drilling of deeper wells, has been required. Impacts subsided after stabilization of the cone-of-depression.

At the Berne field, two 36-in diameter, gravel-packed wells are screened in sand, relatively high in the valley fill. However, formations above included cobbly to bouldery gravels. Aquifer transmissivity, based on performance of these wells, is about 176,000 g/d/ft (2.5×10^{-2} m^2/s). Conductivity is estimated as about 680 ft/d (2.4×10^{-3} m/s).

Deltaic or subaqueous fan aquifers embedded in lacustrine sequences

Sands and coarse-grained gravels are locally embedded within fine-grained, lacustrine sediments of the La Fontaine Segment of the Marion Valley Section, in Wabash and Grant Counties (Fig. 2a). These are interpreted as subaqueous fan and fan-channel deposits of restricted lateral distribution, which reflect proglacial discharge funnelled into the valley through a northern tributary. (These aquifer units are a primary component of up-valley and down-valley lobate, aggregate thickness contours, area D, Fig. 1; also DNR 11, 12, 17, Fig. 6C of Bleuer, this volume.) The specific influx source just north of Marion is probably a single, narrow notch cut in the edge of the ~700-ft flanking surface, as defined by the 650-ft contours of Bruns and others (1985). These aquifers are very complex and discontinuous, yet their proximal parts (the equivalent of the Blackford Member western ice-proximal facies) are probably stacked to the greatest aggregate thickness in areas nearest the points of influx, but north of the valley proper.

In the Marion north well field (Fig. 2a), embedded intralacustrine aquifers are part of a subsidiary aquifer complex associated with the regional capping aquifer that already has been described. These lacustrine aquifer bodies have been characterized by drillers as complex and unpredictable in position, grain size, and vertical or horizontal trend, as is consistent with an interpretation as subaqueous fan, fan-channel, and chute environments. The site consultant's mapping has suggested that these bodies are very narrow and oriented generally northwest-southeast. Thus, although confined entirely within the fine-grained, lacustrine sediments, interpretation suggests that these aquifers must be interconnected hydraulically with each other, as well as with the valley-capping aquifer, particularly in the direction of source; this interpretation is substantiated by interaction between wells in separate aquifers during pumping tests (Stremmel and Hill, 1985). Available pumping data suggest that transmissivities of these lower aquifers range from 24,000 to 103,600 g/d/ft (3.5×10^{-3} m^2/s to 1.5×10^{-2} m^2/s), well within the context of the 85,000 g/d/ft regional average already cited (Stremmel and Hill, 1985).

Similar, but finer grained sands and silty sands are common distal elements of the eastern, ice-proximal facies of the Blackford Member within the upper Geneva Segment of the Marion Valley Section. In the vicinity of Berne, such units are embedded as lenses(?) within fine-grained lacustrine sediments. (The unit is a contributor to aggregate thickness contours, area D, Fig. 1; see also DNR 14, 19, Figs. 6D and 7 of Bleuer, this volume.) If interpreted correctly, these units are expected to vary considerably, pinch out distally, and expand and connect proximally, that is, eastward up the trunk valley and/or northward, up the tributary.

SUMMARY

The Marion-Mahomet trunk valley fill of the Lafayette Bedrock Valley System contains a great variety of glacigenic aquifer facies. These range from stacked, valley-filling masses to thin, lenticular, irregularly distributed, granular, water-bearing intratill units. The deep aquifer in many sectors is confined or semiconfined, being totally saturated but sandwiched between aquitards. Elsewhere, as in areas of crossing of the Marion-Mahomet valley and the modern Maumee-Wabash Trough (Wabash River Valley), thick, superimposed bodies of sand and/or gravel extend from the ground surface to well below the water table, a truly unconfined condition.

The significant variations in the valley-fill aquifers directly relate to a history of incision and infilling that involved differently sourced ice sheets acting in temporal discord. In parts of the valley, there was uninterrupted deposition of granular, valley-rain outwash and braid channel sediments by free-flowing meltwaters. Elsewhere, local deposition was dominated by emplacement of lacustrine clays containing local, embedded mud flows and delta or fan-delta assemblages. In general, "west is best, east is least"; that is, from about Peru-Logansport westward toward Illinois the valley-fill aquifer is thickest, most regular in character, and highly productive, whereas eastward toward Ohio productivity diminishes, and size and continuity of aquifers is less predictable because of a greater complexity of glacial history and deposits.

ACKNOWLEDGMENTS

We are indebted to representatives of the several municipalities mentioned for information and access to test holes and/or production data. Many drillers and consultants have provided important background information over a period of years; information and access to holes specifically discussed herein has been provided by Holt Bros. Drilling, Inc. (Oxford tests), Dr. Darrell I. Leap (Lafayette wells), Ned and Rick Ortman and Jerry Hill of Ortman Drilling, Inc. (DNR, U.S. Geological Survey–Logansport, Marion north well field and other areas), Marsha Taylor of Taylor and Associates (Marion north well field), and Bill Guy of Layne-Northern, Inc. (Berne well field).

REFERENCES CITED

Bleuer, N. K., 1989, Historical and geomorphic concepts of the Lafayette Bedrock Valley System (so-called Teays Valley) in Indiana: Indiana Geological Survey Special Report 46, 11 p.

Bruns, T. M., and Steen, W. J., 1980, Hydrogeology and ground-water availability in the preglacial Teays Valley, west-central Indiana: Geological Society of America Abstracts with Programs, v. 12, p. 221.

—— , 1983, Hydrogeology of the Teays valley fill, north-central Indiana: Geological Society of America Abstracts with Programs, v. 15, p. 535.

Bruns, T. M., Logan, S. M., and Steen, W. J., 1985, Maps showing bedrock topography of the Teays Valley (eastern, central, and western parts), north-central Indiana: Indiana Geological Survey Miscellaneous Maps 42, 43, 44.

Capps, S. R., 1910, The underground waters of north-central Indiana: U.S. Geological Survey Water-Supply Paper 254, 279 p.

Fraser, G. S., and Bleuer, N. K., 1988, Sedimentological consequences of two floods of extreme magnitude in the late Wisconsinan Wabash Valley, *in* Clifton, H. E., ed., Sedimentologic consequences of convulsive geologic events: Geological Society of America Special Paper 229, p. 111–125.

Gillies, D. C., 1981, Ground-water potential of the glacial deposits near Logansport, Cass County, Indiana: U.S. Geological Survey Water-Resources Investigations 81-7, 98 p.

Harvey, E. P., 1990, A hydrochemical investigation of the aquifer system near West Lafayette, Indiana [M.S. thesis]: West Lafayette, Indiana, Purdue University, 319 p.

Maarouf, A. M., and Melhorn, W. N., 1975, Hydrogeology of glacial deposits in Tippecanoe County, Indiana: Purdue University Water Resources Research Center Technical Report 61, 107 p.

Noel, S. D., 1978, Subsurface stratigraphy and water resources of Cass County, Indiana: Purdue University Water Resources Research Center Technical Report 102, 94 p.

Norris, S. E., and Spicer, H. C., 1958, Geological and geophysical study of the preglacial Teays Valley in west-central Ohio: U.S. Geological Survey Water-Supply Paper 1460-E, p. 199–232.

Pohlmann, K. F., 1987, An investigation of the ground-water resources in the Wabash Valley glacial deposits near West Lafayette, Indiana [M.S. thesis]: West Lafayette, Indiana, Purdue University, 147 p.

Rosenshein, J. S., 1958, Ground-water resources of Tippecanoe County, Indiana: Indiana Division of Water Resources Bulletin 8, 38 p.

Stephenson, D. A., 1967, Hydrogeology of glacial deposits of the Mahomet Bedrock Valley in east-central Illinois: Illinois State Geological Survey Circular 409, 51 p.

Stremmel and Hill, 1985, Report and evaluation of proposed north well field study area: Unpublished report(s) for Marion Municipal Water Works.

Thornbury, W. D., 1969, Principles of Geomorphology, 2nd ed.: New York, John Wiley & Sons, 594 p.

Thornbury, W. D., and Deane, H. L., 1955, The geology of Miami County, Indiana: Indiana Geological Survey Bulletin 8, 49 p.

Wayne, W. J., 1956, Thickness of drift and bedrock physiography of Indiana north of the Wisconsin glacial boundary: Indiana Geological Survey Report of Progress 7, 70 p.

Wayne, W. J., and Thornbury, W. D., 1951, Glacial geology of Wabash County, Indiana: Indiana Geological Survey Bulletin 5, 39 p.

Weinreb, G., 1987, A hydrogeologic investigation of thermal contamination within the Wabash Valley outwash aquifer [M.S. thesis]: West Lafayette, Indiana, Purdue University, 170 p.

MANUSCRIPT ACCEPTED BY THE SOCIETY JULY 2, 1990

Printed in U.S.A.

Mahomet Bedrock Valley in east-central Illinois; Topography, glacial drift stratigraphy, and hydrogeology

John P. Kempton
Illinois State Geological Survey, Natural Resources Building, 615 East Peabody Drive, Champaign, Illinois 61820
W. Hilton Johnson
Department of Geology, Natural History Building, University of Illinois, 1301 West Green Street, Urbana, Illinois 61801
Paul C. Heigold and Keros Cartwright
Illinois State Geological Survey, Natural Resources Building, 615 East Peabody Drive, Champaign, Illinois 61820

ABSTRACT

The buried Mahomet Valley in east-central Illinois is a complex lowland carved into the surface of Pennsylvanian and older rocks. It consists of a deep channel throughout most of its length and contains numerous benches below erosional remnant hills, suggesting several cycles of early to middle Quaternary erosion. Recent local and regional studies utilizing existing borehole data, including down-hole geophysical logs and seismic profiles, have provided new insights into the valley's configuration. The usual techniques for interpreting shallow seismic refraction and reflection data are complicated by a seismic velocity inversion in the Quaternary sediments filling the deeper parts of the valley.

Based on the 500-ft (152 m) elevation contour to define the upper limit of the Mahomet Valley Lowland, the valley is about 8 mi (13 km) wide at the Illinois/Indiana border. Westward, the lowland widens to as much as 18 mi (29 km) in Ford County, where a major tributary enters from the north. Here, several ridges rise above a broad bench ranging from 400 to 450 ft (121 to 137 m) in elevation. Sparse well data and some seismic profiling suggest a deep channel at or slightly below an elevation of 350 ft (106 m) in northeastern Ford and northwestern Vermilion Counties. This deep channel extends southwestward where it passes under the villages of Mahomet (Champaign County) and Monticello (Piatt County) and then westward to just east of Clinton (De Witt County). Between Mahomet and Clinton, where there are also several isolated bedrock hills that rise to elevations as much as 500 ft (152 m), the intermediate bench lowers to 350 to 400 ft (106 to 121 m). The valley narrows to an average of 14 mi (22 km) wide between Monticello and its confluence with the Mackinaw Valley segment of the Ancient Mississippi Bedrock Valley in southwestern Tazewell County. Southeast of Clinton, a narrow bedrock "diversion" channel (Kenney Valley) provides a nearly straight connection between the Mahomet Valley and the Ancient Mississippi Valley below the confluence with the Mahomet.

A complex Quaternary history has been established for the Mahomet Bedrock Valley, but as yet no evidence has been found for late Tertiary or preglacial alluvial deposits. Deposits filling the valley include the widespread Mahomet Sand Member, as much as 200 ft (60 m) thick, locally overlying or interbedded with tills of the Banner Formation (pre-Illinoian). Above this succession are Glasford Formation (Illinoian) and Wedron Formation (Wisconsinan) tills and associated deposits. The varied nature of the bedrock valley topography, the scattered presence of till-like material on bedrock hills underlying the Mahomet Sand, and the presence of lower Banner Formation till interbedded with the Mahomet Sand suggest several episodes of valley erosion and glacial deposition during a long pre-Illinoian history.

The deepest bedrock channel probably originated before deposition of the Mahomet Sand but postdates at least one early Quaternary glaciation. Near the Indiana border, the uppermost surface of the Mahomet Sand, at elevation 560 ft (170 m), appears locally eroded, forming broad terraces that continue down valley at progressively lower elevations. The surface of the Banner Formation forms a broad sag over much of the valley and rises slightly over the uplands. Deposits of the Glasford Formation form the upper fill in the valley and include a significant outwash related to the Vandalia Till Member. The Sangamon Soil, developed in the Glasford, together with the overlying Roxana Silt and Robein Silt, locally forms an important subsurface marker. The topographic expression of this pre-Woodfordian surface shows no evidence of the Mahomet Valley; it was completely buried by the end of the Illinoian.

Aquifers associated with the Mahomet Bedrock Valley and the Ancient Mississippi Bedrock Valley to the west are the only highly productive, nonalluvial sand and gravel aquifers in the southern three-fourths of Illinois. The aquifers associated with the buried Mahomet Valley provide the only large source of irrigation, industrial, and municipal supplies of groundwater in east-central Illinois; 40 municipalities and water districts are currently obtaining groundwater from these aquifers.

The largest groundwater withdrawals occur in the Champaign-Urbana area, averaging 17×10^6 gal (64×10^6 l/day. Total groundwater withdrawals from the valley are estimated to be at least 42×10^6 gal (16×10^7 l)/day. The coefficients of storage for the Mahomet Sand range from 2×10^{-5} to 2×10^{-3}, with hydraulic conductivities and transmissivities up to 4,237 gpd/ft^2 (2×10^{-3} m/s), and 510,000 gpd/ft (5×10^{-2} m^2/s), respectively; for the Glasford sand the coefficients range from 1×10^{-5} to 8×10^{-2}, with hydraulic conductivities and transmissivities up to 4,660 gpd/ft^2 (2×10^{-3} m/s), and 233,000 gpd/ft (2×10^{-2} m^2/s), respectively. Coefficients of vertical hydraulic conductivity of the confining beds range from 2.12×10^{-3} to 0.4 gpd/ft^2 (1×10^{-9} to 2×10^{-7} m/s).

INTRODUCTION

Background and purpose

The Mahomet Bedrock Valley in east-central Illinois (Figs. 1, 2) has been the subject of much interest since it was first defined by Leland Horberg in 1945. This interest stems primarily from the importance of sand and gravel aquifers that are contained within the valley and the history of erosion and deposition during and prior to Quaternary continental glaciation as recorded by the valley and its fill. There has also been considerable interest by the general public, as well as by geologists, in the occurrence of such buried valleys, and in particular in the Mahomet, as a possible continuation of the "Teays" drainage system, thought by some to have originated in southeastern West Virginia in late Tertiary or early Quaternary time. The idea of a major drainage system ("river"), now buried, has been the subject of many popular articles.

In Illinois, no comprehensive study of the Mahomet Valley System has been undertaken since the work by Horberg in the mid-to late 1940s. However, during the more than 40 years since Horberg's original work, much new subsurface data have accumulated, and numerous local and regional groundwater and related studies have yielded new insights into the configuration and history of the valley. In addition, the use of more sophisticated techniques, both geophysical and drilling and sampling, has improved the quality of data available. With new data and recent studies in Indiana, Ohio, and adjacent states of the bedrock topography and fill of the valleys associated with the Teays System, renewed interest has developed on the relation of the Mahomet Valley to these bedrock valleys. There is also an ever-increasing need in Illinois to better understand the distribution of aquifers and the groundwater resource potential within the Mahomet Valley.

Whereas work on updating the bedrock topography of Illinois is ongoing, study of the Mahomet Valley was accelerated and this status report completed because of the importance of the aquifers contained within the Mahomet Valley. There was also the desire to tie together the numerous fragmented studies of portions of the valley, evaluate significant new data—both on the configuration of the bedrock valley and the stratigraphic record of the fill it contains—and to complete an overall update of Horberg's work.

The Mahomet Bedrock Valley was the major drainageway

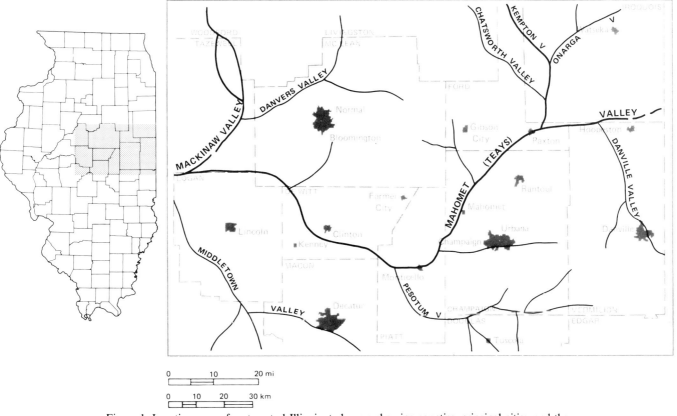

Figure 1. Location map of east-central Illinois study area showing counties, principal cities, and the thalweg of Horberg's Mahomet Valley and tributaries (Horberg, 1950, Plate 2).

in east-central Illinois (Fig. 1) during the early and middle Pleistocene, and possibly earlier. It carried runoff, including large volumes of glacial meltwater, from eastern and northern Illinois, at least parts of northern Indiana, and possibly areas farther east. An understanding of the valley is dependent on knowledge of its configuration and character, and of the deposits contained within it and its tributary valleys. From these data, inferences with regard to the history of the valley can be deduced, and areas where more and/or better basic subsurface data are needed can be identified. The latter are essential for further development of the drainage and glacial history and for groundwater exploration and water resource evaluation.

Previous studies

Horberg (1945) first mapped and delineated the Mahomet Valley and concurrently described the general sequence of deposits filling the valley (Fig. 2). Later, Horberg (1950, 1953) described more details of the stratigraphic sequence and outlined the general history of the bedrock surface and associated valleys. Horberg's work was outstanding, and his basic stratigraphic framework and interpretations differ only moderately from our current concepts.

Horberg (1945, 1950, 1953) interpreted the deep valley to be preglacial in origin, and a thick sand and gravel fill in the valley to be either pro-Nebraskan or pro-Kansan. These deposits were named the Mahomet Sand and were shown to overlie bedrock and to occur below "Kansan" drift. Horberg mapped buried Yarmouth and Sangamon surfaces, which suggested that the Mahomet Valley was filled with drift and essentially abandoned after Kansan glaciation. Locally, sags in the Yarmouth surface coincided with the bedrock valley and were related to drainage and valley enlargement during the Yarmouthian.

Subsequent work in and near the Mahomet Valley has been concerned primarily with groundwater exploration and evaluation. These studies utilized the stratigraphic approach developed by Horberg and expanded the subsurface data base. Many were of limited scope related to local municipal or farm water supplies and are available as unpublished reports at the Illinois State Geological and Water Surveys. Other local studies were published or were more regional in scope and include work by Foster and Buhle (1951), Foster (1952, 1953), and Selkregg and Kempton (1958).

Stephenson (1967) used a hydrostratigraphic approach in a study of Mahomet Valley aquifers, but due to absence of more definitive stratigraphic control, his resulting stratigraphic framework was similar to that of Horberg. He also developed maps of the configuration of the Yarmouth and Sangamon surfaces that generally agreed with Horberg's, except he showed a better defined valley over the Mahomet Bedrock Valley on the Yarmouth surface map.

Data available

Significant improvements in the quantity and quality of data since Horberg's work are exemplified by the results of recent studies. These studied have utilized new data to prepare much more detailed and reliable maps of the bedrock topography and some glacial drift units. They have utilized the continually growing collection of subsurface data, including water well logs and samples, test-hole logs and samples, drift cores, and downhole geophysical logs (mostly gamma, resistivity, and spontaneous potential (SP). For some commercial exploratory drilling for drift aquifers and engineering characterization, core sets are available. Drift cores are also available from a controlled drilling program completed for a water-resources study (Kempton and others, 1982) mainly south of the Mahomet Bedrock Valley.

A considerable number of commercial downhole geophysical logs into bedrock are available for the region, including about 200 that are in or immediately adjacent to the Mahomet Bedrock Valley. These logs have been utilized in the more recent reports to define the bedrock topography and to differentiate principal lithic units within the drift. They have also been used to tie between local map areas and to provide the bulk of the definitive data for supplemental mapping of the bedrock surface outside previously mapped areas.

Surface geophysical methods, principally resistivity and seismic refraction surveys, have been used as an aid in locating sand and gravel aquifers and to provide data on the depth to bedrock. Gravity data have also been used (Heigold and others, 1964). Even though there is in general a density contrast between glacial drift and most of the bedrock of the region, the location of large bedrock valleys often coincides with highs on residual gravity maps.

Geologic data and maps. Records of water wells probably make up more than 90 percent of the subsurface information available throughout the Mahomet Valley region, a total of approximately 4,500 well locations. It is estimated that less than 10 percent of these penetrate bedrock. Records for nearly 350 commercial test holes and wells (structure, coal and oil tests, gas storage, and oil wells) are also available and are relatively well distributed throughout the region, although sparse in southern McLean County. A relatively few engineering (foundation-exploration) borings for bridges, dams, water-supply tests, miscellaneous structures, and some controlled drilling using similar drilling and core sampling methods are available; only about 20 of these reach bedrock. Records from a number of municipal and industrial wells and a few domestic and farm wells include sample cuttings supplied by the drilling contractors. Nearly all of the commercial test holes and a number of water wells have

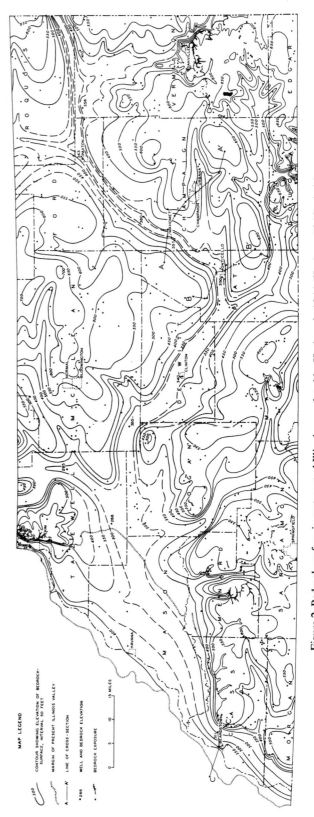

Figure 2. Bedrock surface in east-central Illinois completed by Horberg in 1944 (Horberg, 1945). A-A', B-B', and C-C' are lines of cross section shown in Horberg, 1945.

been geophysically logged (usually natural gamma and/or resistivity—SP).

We estimate that there has been a 75 percent increase in data available on the bedrock surface since Horberg's work in the mid- to late 1940s, of which a good percentage has significantly added to our understanding of the configuration of the Mahomet Valley. Although driller's logs, sample cuttings, and geophysical logs have provided good information on the general drift character, cores (split-spoon samples) from key test holes and water wells have provided the representative samples from which the most meaningful stratigraphic interpretations and correlations have been made.

Maps in this report are the result of compilation of all existing data and interpretations available through 1983, both published and unpublished, along with reconnaissance studies of northern Champaign, Ford, and Iroquois Counties, small portions of southern and southeastern McLean County, and northern Piatt County. Portions of unpublished maps of the bedrock topography of McLean County (Nelson, 1981) and a large portion of east-central Illinois (Cartwright, 1972) are also incorporated. The maps that were prepared for previous or concurrent studies and compiled for this study of the entire Mahomet Valley region were based principally on the subsurface data described above. These include maps of the bedrock topography and Mahomet Sand distribution in western McLean and eastern Tazewell Counties (Richards and Visocky, 1982), DeWitt County (Hunt and Kempton, 1977), the southern portion of east-central Illinois (Kempton and others, 1982), and northern Vermilion County (Kempton and others, 1981).

Additional subsurface data were compiled to provide a complete and consistent interpretation of the bedrock topography and aquifer distribution of the entire Mahomet Valley area, and provide a tie between previously mapped areas. Most of the available data for the more recent studies had been plotted on 15-min topographic maps for the entire area except for the northeastern portion (extreme Livingston, northern Ford, and Iroquois Counties). A review was made of the well-log files at the ISGS to pick up any new data in the tie-in areas, and all wells penetrating bedrock were plotted for the northern area and checked against Cartwright's (1972) map.

Surface geophysical data. Surface geophysical methods have been used for many years throughout Illinois to prospect for sand and gravel aquifers and determine bedrock topography. These methods have been principally resistivity and seismic refraction surveys, although gravity methods have also been employed (Heigold and others, 1964; McGinnis and others, 1963). Recent seismic refraction surveys in combination with borehole data have been quite successful in adding considerable detail to the definition of buried bedrock valleys (McGinnis and Heigold, 1974; Heigold and Ringler, 1979; and earlier, by Bredehoeft, 1957). In addition, a number of scattered seismic surveys profiling bedrock valleys, such as the Paw Paw Valley (Kempton and Reed, 1973) have supplemented ongoing studies.

A supplemental refraction survey was made in the Mahomet

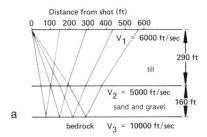

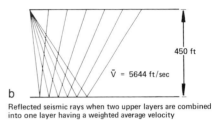

Figure 3. a, Schematic diagram depicting *reflected* seismic rays and layering parameters (velocities and thicknesses) similar to those in the Mahomet Bedrock Valley. b, Seismic refraction travel time plot for the layering parameters shown in Figure 3a. A segment corresponding to V_2 = 5,000 ft/s is missing—the hidden layer.

Valley in northern Vermilion and Ford Counties to attempt to locate the thalweg of the Mahomet Valley and determine its elevation. Prior to this survey, there had been a lack of progress with seismic refraction surveys, particularly in the Mahomet Valley area, because of the hidden layer problem encountered. The refraction method assumes increasing velocities with depth, e.g., till or sand and gravel over bedrock. However, within the Mahomet Valley, the common sequence is thick till overlying an equally thick, lower velocity sand and gravel that in turn lies on bedrock. Figure 3a is a schematic diagram depicting refracted seismic rays and layering parameters (velocities and thicknesses) similar to those in the Mahomet Valley. The velocity inversion in the unconsolidated sediments results in the hidden layer problem of the seismic refraction method. Figure 3b shows a seismic refraction traveltime plot for the layering parameters shown in Figure 3a. A segment corresponding to V_2 = 5,000 ft/s is missing—the hidden layer. Calculation of depth to bedrock using this plot yields 497 ft (150 m), much deeper than the real value of 450 ft (136 m). This discrepancy makes the seismic refraction method of questionable value in this situation. With the presence of a hidden layer the seismic refraction methods can reliably locate the thalweg but cannot accurately determine depth to bedrock.

As an alternative to seismic refraction methods, an attempt was made to use reflection methods. Figure 4a is a schematic diagram depicting reflected seismic rays and layering parameters as in the Mahomet Valley; Figure 4b shows reflected seismic rays

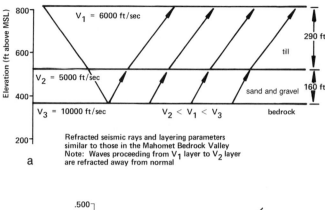

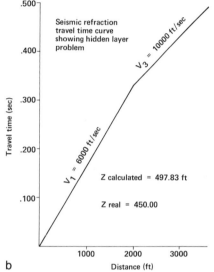

Figure 4. a, Schematic diagram depicting *reflected* seismic rays and layering parameters similar to those in the Mahomet Valley. b, Reflected seismic rays when two upper layers are combined into one layer having a weighted average velocity.

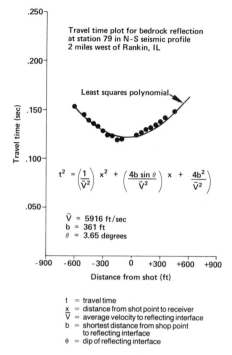

Figure 5. Typical bedrock reflection travel time plot from the Mahomet Bedrock Valley. Assuming valley-fill sediments are one layer with average velocity $\bar{v}$, a least squares polynomial fitted to the time-distance data yields average velocity, depth to bedrock, and bedrock (surface) slope.

when the two upper layers are combined into one layer having a weighted average velocity.

A hypothetical bedrock reflection traveltime plot, typical of the Mahomet Valley, was prepared (Fig. 5). Assuming that the unconsolidated sediments are one layer of average velocity, $\bar{V}$, and the bedrock surface is planar, a least squares polynomial fitted to the time-distance data yields average velocity, depth to bedrock, and bedrock dip. This technique has been found to provide depths to bedrock accurate to ±10 ft, and was applied to seismic profiling at three traverses across the Mahomet Valley. Although additional information on downhole velocities in the various drift materials would further increase the accuracy of this method, shallow seismic reflection is the preferred geophysical technique to map the bedrock surface.

Electrical earth resistivity (EER) methods have been used extensively in Illinois for more than 50 yr (Buhle and Brueckmann, 1964), both in scattered-site groundwater exploration and in more extensive surveys for mapping aquifer distribution, as was undertaken for northern Vermilion County (Kempton and others, 1981). While often effective in defining the extent and thickness of shallow sand and gravel deposits, its effectiveness in determining depths to bedrock is limited due to similar resistivities of contrasting lithologies at the bedrock surface (e.g., sand over carbonate rock or till over shale). The existing bedrock geologic map (Fig. 6) has been a guide for identifying bedrock lithologies for both seismic and resistivity work.

Key drift stratigraphic sections. As in work with surface exposures, availability of definitive subsurface data is often a matter of chance. Basic descriptions of representative boring samples must provide the principal data for defining or recognizing stratigraphic units in the subsurface. Until fairly recently, the only data that provided some information on the subsurface drift stratigraphy were the logs and drill cuttings from water wells that were submitted by drilling contractors. The distribution of sample sets was random and represented materials from only a very small fraction of the wells drilled in the region each year. The quality of the samples varied, partly due to drilling methods (e.g., cable tool versus rotary) and particularly due to care taken during sample collection. Normally, only cuttings from large-capacity commercial, industrial, irrigation, and municipal wells and test holes are available, although samples from domestic and farm wells and test holes are occasionally submitted.

The availability and value of cores of the glacial drift taken

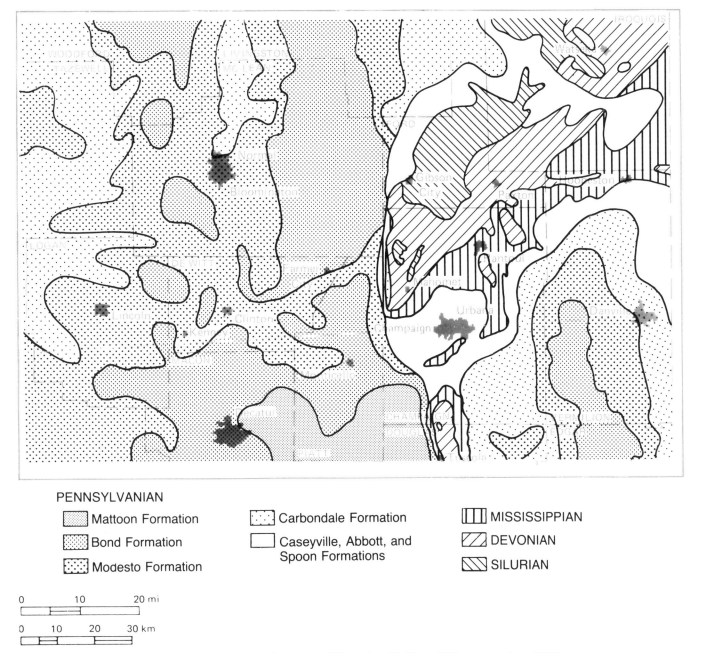

Figure 6. Bedrock geology of east-central Illinois (modified from Willman and others, 1967).

in conjunction with engineering projects was recognized more than 25 yr ago. Soon their importance was also recognized as a key to understanding regional glacial drift stratigraphy and properties (Kempton and Hackett, 1962, 1968). It is this type of information, along with those stratigraphic sections exposed in strip mines and other exposures over the adjacent bedrock uplands, that has provided the basis for the drift stratigraphy described herein. The specific projects that have provided the key stratigraphic data utilized in this chapter, described subsequently, are listed below:

- Foundation borings for Lake Evergreen Dam (for City of Bloomington), 1966
- Initial engineering investigations for Clinton Lake Dam (Illinois Power Company), 1968
- Pilot and subsequent test holes for Champaign-Urbana west well field expansion (Northern Illinois Water Corp), 1968

- Pilot and subsequent test holes for new Town of Normal well field, 1969
- Controlled drilling for aquifer study in southern part of east-central Illinois (providing five test holes in the uplands just south of Mahomet Valley), 1978
- Test holes for irrigation supply (Cole Farm); split-spoon samples of lower section, 1979
- Sample cuttings from city of Weldon Test Hole 4-62 (cable tools), 1962, studied in 1970
- Sample cuttings from test well for Burrill Hall, University of Illinois, 1955, studied in 1968
- Water well, Village of Hoopeston
- Test hole near Ambia, Indiana, part of controlled drilling for study of "Teays" Valley in Indiana— eastern tie-in hole (Bleuer, this volume)

The principal outcrops that provided tie-in of the stratigraphic sequence came mostly from strip-mine exposures from the bedrock uplands in Vermilion County (Johnson and others, 1971; Johnson and others, 1972). Correlations between control borings and the known drift sequence were based on the presence of key beds, buried soils, and on the color, clay and carbonate mineralogy, and texture of the till units.

Numerous water-well sample sets and driller's logs and downhole geophysical logs were selected to draw cross sections to tie in the key control holes. Once the stratigraphic sequence was identified from key borings and sections, correlations could be made by identifying general sequences between control borings on the basis of general lithologic descriptions and geophysical log configurations.

BEDROCK TOPOGRAPHY OF THE MAHOMET BEDROCK VALLEY REGION

The current revision (Figs. 7, 8) of Horberg's map provides a new perspective on the configuration of the bedrock topography, confirming the essence of Horberg's map but at the same time adding important changes in detail. From a regional perspective the broad pattern of uplands and major valleys has not changed, but the shape and width of the Mahomet Valley has, and tributary valleys are locally added or shifted as new data have suggested different positions.

The description and interpretation of the Mahomet Bedrock Valley presented here, the special features of the valley, and the nature of its tributaries and related channels set the stage for discussion of the valley-fill stratigraphy and an updated interpretation of its development and history. The current map also provides a basis for further work and interpretations for applied studies (e.g., groundwater resource evaluation).

Regional bedrock characteristics

To understand the location, characteristics, and development of the Mahomet Bedrock Valley as currently mapped, it must be studied in relation to its regional setting. This setting includes the bedrock lithology, structure and topography, its preglacial development, and the processes and effects of glaciation. Although some of the discussion on early valley development is speculative or interpretive, there are some factual data that provide a substantive basis for discussion.

Horberg and Anderson (1956) suggested that Pleistocene glacial lobes were closely controlled by bedrock lowlands. They prepared a map of the bedrock topography of the central United States that suggested to them that the buried topography of the lowlands (elevations generally below 750 ft, or 229 m) is due mainly to stream erosion and associated processes. They observed that, throughout most of the midcontinent region, major lowlands and uplands conform very closely to bedrock lithology and structure.

Bedrock formations in east-central Illinois (Fig. 6) consist of a succession of sedimentary rocks several thousand feet thick, including sandstone, limestone, dolomite, shale, and coal. These rocks were warped and tilted to form the Illinois Basin centered in southeastern Illinois and an arch-like structure, the La Salle Anticlinal Belt, which trends north-south through the approximate center of the study region.

Whereas the older, generally deeper rocks of east-central Illinois are, for the most part, composed of limestone, dolomite, and sandstone, the younger Pennsylvanian rocks that are at or within a few hundred feet of the bedrock surface are composed largely of shale interbedded with relatively thin layers of sandstone, limestone, and coal. Only along the trace of the La Salle Anticlinal Belt do the older rocks reach near or to the bedrock surface.

A study of Horberg's map, as revised and summarized by Willman and Frye (1970, Fig. 4, p. 19), shows that the Mahomet Valley lies within an area where the bedrock is generally below 600 ft (183 m) in elevation, with a discontinuous rim of bedrock at elevations above 600 ft (183 m) both to the north and south. The rocks that underlie this rim are composed predominantly of Pennsylvanian shales with some sandstones and some significant regionally persistent limestones (Millersville and Shoal Creek of the Bond Formation) mainly west of the La Salle Anticlinal Belt. However, the northeastern portion of this rim is composed of Silurian dolomite (Fig. 6). Devonian and Mississippian rocks underlie the lower portions of the Mahomet Valley in northern Champaign, southeastern Ford, northern Vermilion, and southern Iroquois Counties. Because of pre-Pennsylvanian erosion, Pennsylvanian rocks lap directly onto the Silurian in Ford and Iroquois Counties and progressively younger rocks to the southeast in Champaign and Vermilion Counties (Fig. 6; Willman and others, 1967).

Definition of the Mahomet Valley

Position and general features. The general location of the Mahomet Valley has been changed only in detail by the revised mapping presented in this report (Fig. 7). Although the revised map generally follows Horberg's map, it reflects more specifically

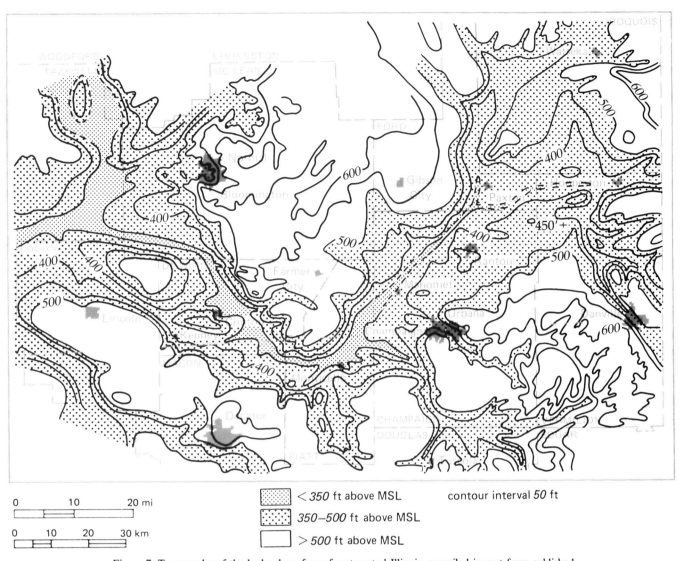

Figure 7. Topography of the bedrock surface of east-central Illinois, compiled in part from published maps in Hunt and Kempton (1977), Kempton and others (1981, 1982), and others, as well as the unpublished maps of Cartwright (1972) and of Reinertson and others (1977).

Willman and Frye's (1970) map in some areas and provides considerably more detail, which changes the overall impact, appearance, and configuration of the valley from that of previous maps. The thalweg of the Mahomet Valley and its principal tributaries, including the Kenney Valley named herein, are shown in Figure 8.

There are several significant differences between the current map and previous maps of the Mahomet Valley. Our mapping indicates that: (1) the main channel of the Mahomet Valley is located northeast of Clinton in central DeWitt County (Heigold and others, 1964; Stephenson, 1967; Hunt and Kempton, 1977); (2) the Pesotum Valley enters the Mahomet Valley in east-central Piatt County (Heigold and others, 1964; Kempton and others, 1982); (3) there is no evidence of a channel below 300 ft (92 m)

elevation (Stephenson, 1967); and (4) there is no evidence for the Harris Valley, an adjacent valley mapped by Heigold and others (1964) to the northwest of the Mahomet Valley in eastern DeWitt County and northern Piatt County (Hunt and Kempton, 1977). Most of these changes had been accepted by Willman and Frye (1970) in their generalized map of the bedrock topography in Illinois.

The current map (Fig. 7) shows the Mahomet Valley and its major tributaries generally to be outlined by the 500-ft (153 m) elevation contour. This area is referred to as the Mahomet Valley Lowland. From the Indiana state line westward, the lowland broadens from about 8 mi (13 km) in width to approximately 20 mi (32 km). This area includes the confluence of the Danville Valley from the south, in northern Vermilion County, and a

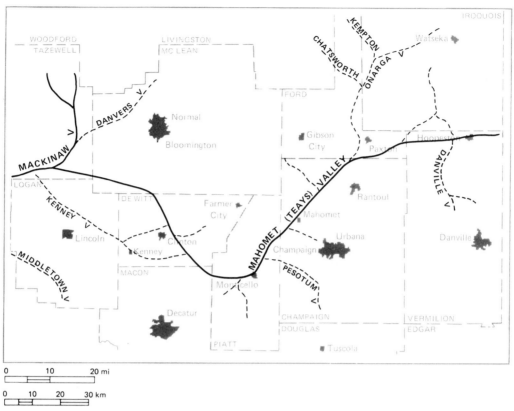

Figure 8. Thalweg of principal bedrock valleys as presented in this report. Solid lines are the main channels.

major valley, the Onarga Valley, from the north, near Paxton in Ford County. Southwestward, the lowland narrows progressively across northwestern Champaign County into Piatt County, where it reaches a minimum width of about 8 mi just west of Monticello and then turns rather sharply northwestward, almost 90°.

From western Piatt County the lowland then widens again through northeastern Macon County and southern DeWitt County to just north of Clinton, where it attains a width of more than 15 mi (24 km). At that point an elongate bedrock ridge, slightly above elevation 500 ft (152 m), separates the "main" channel from a narrower channel to the southwest. The "main" channel turns nearly west in southwestern McLean County, then opens into the wide lowland of the Mackinaw Valley segment of the "Ancient Mississippi Bedrock Valley" at the southeastern corner of Tazewell County. The narrower channel, about 5 mi (8 km) wide, beginning at the Village of Kenney in southwestern DeWitt County, is here named the Kenney Valley (Fig. 8). It trends northwestward across northeastern Logan County, joining the Mackinaw Valley about 12 mi west of the confluence of the "main" Mahomet Channel.

Within the Mahomet Valley Lowland, there are a number of noteworthy features: (1) steeper slopes along the rim of the lowland where it is wide—with steeper slopes in general along the narrower portions of the lowland; (2) broad benches north-east of Paxton, mainly in Iroquois County at an elevation just below 500 ft (152 m), and a lower bench, between elevation 400 and 450 ft (122 and 138 m), extending from just northeast of Champaign-Urbana to north-central Vermilion County, with possible remnants along the remainder of the lowland; (3) the predominant northwest-southeast and northeast-southwest trend to the lowland and most tributaries; (4) a broad inner valley below elevation 400 ft (122 m) that is persistent and well documented from the Indiana state line westward to the Mackinaw Valley; (5) the lack of substantive evidence of a channel below 300 ft (91 m) in elevation; and (6) only scant information to suggest a continuous channel below 350 ft (107 m) in elevation.

The margins of the Mahomet Valley Lowland are rather sharply defined throughout much of its extent, and verified at numerous localities where closely spaced, reliable well data are available. This is particularly true along the northern sides of the lowland in Piatt and DeWitt Counties, the south side of the lowland just west of the Illinois-Indiana state line and along segments near Champaign-Urbana and in western Piatt County. The area between elevations 500 and 450 ft (152 and 138 m) around the edge of the lowland appears as a relatively narrow rim generally no more than a mile or two wide.

Although the contour interval selected for mapping the bedrock topography may produce somewhat artificial impressions, the 400-ft (122 m) contour also appears to generally mark the upper boundary of another significant feature, a main channel or valley within the Mahomet Valley Lowland. Whereas there are considerable data along the entire length of the valley to verify an inner lowland or valley below 400 ft (122 m), data are too sparse to firmly document a continuous channel below 350 ft (107 m). There are no confirmed data to suggest a channel below elevation of 300 ft (91 m).

Elevations recorded as below 350 ft (107 m) in the eastern segment of the valley occur near the Ford-Vermilion County line (suggested by seismic profiles, Figs. 9, 10), at 336 ft (102 m), 5 mi (8 km) southwest of Paxton (geophysical log), and at 348 ft (106 m) near Paxton in Onarga Valley (water well). In the central segment, low elevations are reported at 331, 318, and 291 ft (101, 97, and 89 m) just southwest of Mahomet (casing records of structure tests), at 338 ft (103 m) in east-central Piatt County (casing record in the Pesotum Valley near its confluence with Mahomet Valley) and at 328 ft (100 m), 5 mi (8 km) east-northeast of Clinton (geophysical log) in the western segments. An elevation of 341 ft (104 m) is reported in the Mackinaw Valley, 6 mi (10 km) west of the confluence with the Mahomet. A few other recorded points are less firm.

Walker and others (1965), studied the groundwater resources of the Havana Lowlands just west of the confluence of the Mahomet with the Mackinaw and reported, in their original maps, a deep channel below elevation 320 ft (98 m). In an addendum, however, the results of a seismic survey were reported with an updated map suggesting the bedrock to be no lower than 350 ft (107 m). Heigold and Ringler (1979), in a seismic refraction survey of the lower Illinois River Valley (Lower Ancient Mississippi Bedrock Valley), showed no elevation below 341 ft (104 m) in their northernmost line, about 70 mi (113 km) down

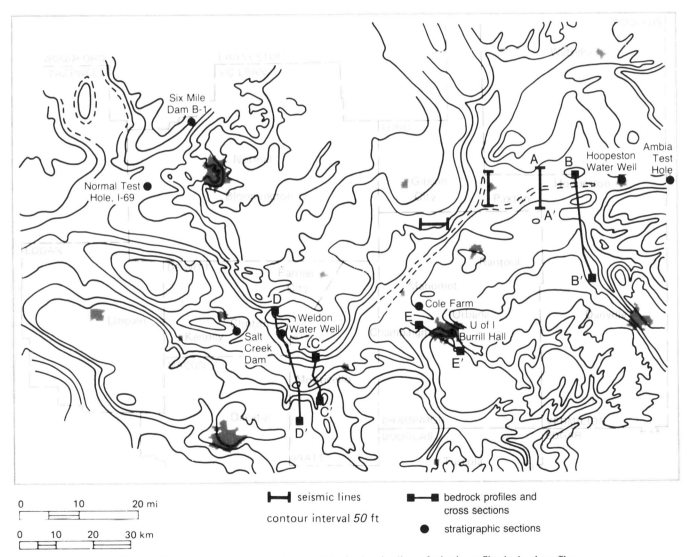

Figure 9. Bedrock topography of east-central Illinois, showing lines of seismic profiles, bedrock profiles, and drift cross sections, and location of selected stratigraphic sections. Data for seismic profiles shown were obtained during 1982–1983 field season.

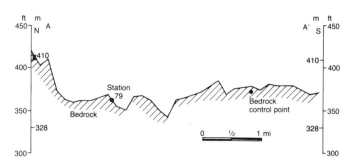

Figure 10. Eastern of two north-south seismic profiles of the bedrock surface (A-A'), near Vermilion-Ford county line (see Fig. 9), determined by least-square polynomial technique.

valley from the Mahomet confluence. They do, however, show two water wells with bedrock at elevations of 330 and 327 ft (101 and 102 m). Farther down valley, lower elevations are recorded. North of the confluence with the Mahomet, a few wells penetrate bedrock in the deeper parts of the Mackinaw Valley, and only one suggests the elevation of 346 ft (105 m); others nearby are above 360 ft (110 m).

Special features. There are a number of features of the Mahomet Valley Lowland that have been highlighted by the current study. These generally can be categorized as related to shape, depth, and orientation, all with implications as to mode of formation and overall characterization, and therefore greater predictability of the Mahomet Valley Lowland, including the sedimentary fill.

The shape of the lowland as presently mapped is considerably different in detail than previously implied. The benches and related hills and the secondary channels are of particular interest. Horberg (1945, 1950) showed a bifurcated valley in the upper part of the Mackinaw Valley in northeastern Tazewell County (Figs. 1, 2). This feature has been the subject of some interest but has generally been treated as a local aberration. The elongate bedrock ridge in northeastern Logan County was noted in Horberg's map and was later shown as separated by a channel by Heigold and others (1964) and later by Stephenson (1967). The former authors also suggested a bedrock hill within the Mahomet Valley in eastern Champaign County, which was not shown by the latter. The presence of "remnant" hills within the Mahomet Valley Lowland surrounded by associated channels has been confirmed and more have been mapped.

Bedrock benches. Among the more interesting features of the Mahomet Valley Lowland are the broad benches (or bedrock terraces), which are particularly well developed in northeastern Champaign County and northern Vermilion County between elevations 400 and 450 ft (122 and 137 m), and in southwestern Iroquois County, between elevations 450 and 500 ft (137 and 152 m). The latter, trending southwest-northeast, about 20 mi (32 km) long and about 8 mi (13 km) wide, appears to be underlain by both Devonian and Pennsylvanian shales and dolomites (Fig. 6), although at the southwestern end a small area above 500 ft (152 m) may be held up by Silurian dolomite along the La Salle Anticlinal Belt. This surface is generally flat to rolling, with elevations probably ranging from 470 to 480 ft (143 to 146 m), although data are sparse.

The even more extensive bench at elevation 400 to 450 ft (122 and 137 m) that lies along the valley lowland between the confluence of the Pesotum Valley to the southwest and the Danville Valley to the northeast is nearly 40 mi (64 km) long and as much as 10 mi (16 km) wide near Rantoul. It appears to be developed across Pennsylvanian shales and possibly Mississippian siltstones. It is characterized by a gently undulating surface with a line of small hills rising to slightly above elevation 450 ft (137 m), the largest documented of which lies under the village of Rantoul. The small hill just to the northwest of Champaign, recognized from samples and logs from a series of irrigation wells, is capped by possible till; it is discussed in more detail below. The line of hills lies close to the deeper valley; some of the lowest elevations of record on this bedrock bench lie close to the lowland walls from southeast of Rantoul to just north of Champaign (404 to 411 ft, or 123 to 125 m), suggesting the presence of a channel (e.g., Fig. 11a).

Remnant hills and channels. The most intriguing of these "hills" and related channels are those associated with the bedrock bench northeast of Champaign, the small remnant in east-central Piatt County, and the large ridge and associated channel (Kenney Valley) to its south in northeastern Logan County. Several of these hills and ridges are elongate and parallel to the valley and thus have a "streamlined" appearance.

The hills that rise above the extensive bench northeast of Champaign fall nearly on a northeast-southwest line from just southwest of Hoopeston in north-central Vermilion County to just northwest of Champaign. Of the four hills mapped (Fig. 7), the one in northern Vermilion County is probably the highest, reaching at least elevation 493 ft (150 m). It is elongated and extends in nearly an east-west direction for nearly 6 mi (10 km), apparently lower, 460 ft (140 m) to the west. The hill underlying Rantoul is irregularly shaped and reaches about elevation 470 ft (143 m); it is probably the largest of the four mapped. Of the other two hills, the one in northeastern Champaign County reaches an elevation of 466 ft (142 m); the other, just northwest of Champaign, reaches 458 (140 m).

Although there is adequate information to assure that the two hills to the northeast are isolated, there are no wells penetrating bedrock between the Rantoul hill and that northwest of Champaign; therefore, it is possible that these two could form a continuous ridge at an elevation about 450 ft (137 m). Also, since few data exist north of Champaign-Urbana to Rantoul, there is even the possibility that much of the area to the north and northwest of Champaign-Urbana is above elevation 450 ft (137 m).

The other feature associated with this bench is an apparent

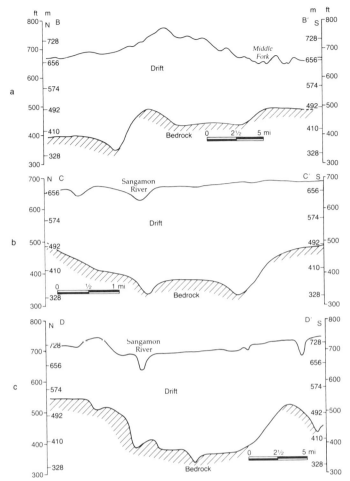

Figure 11. a. Bedrock profile (B-B'), across Mahomet Valley in northern Vermilion County, showing irregularity of bedrock surface. b. Bedrock profile (C-C') across Mahomet Valley in central Piatt County showing two bedrock channels. c. Bedrock profile (D-D') across Mahomet Valley from southeastern DeWitt County to southwestern Piatt County, showing the edge of the remnant hill on the south side of the valley.

channel below elevation 425 ft (130 m) that appears to extend at least discontinuously south or southeast of these hills from the Danville Valley on the northeast, southwestward to just northwest of Champaign, generally paralleling the south side of the lowland. Elevations of this "channel" are as low as 404 ft (123 m). Because control is relatively scattered, the continuity of these lows is still speculative, but is suggestive of at least one or more channel segments. The fact that along the entire length of the bench these "lows" frequently fall between the hills to the northwest and the lowland wall to the southeast lends some support to the existence of a single channel.

A parallel but isolated example of a "remnant" hill within the lowland is that in west-central Piatt County. This hill was delineated by Kempton and others (1982); a portion of their map is reproduced as Figure 12. This hill lies near the south wall of the lowland in contrast to those described farther to the northeast. It reaches an elevation just above 500 ft (152 m), and is capped by the Millersville Limestone. Geophysical logs suggest that the Millersville caps the surface of the bedrock uplands just to the south of the Mahomet Valley Lowland (Clegg, 1972). This hill was "discovered" when a farm family inquired as to why they did not penetrate sand and gravel but hit limestone instead. Field inspection of the well cuttings and a subsquent seismic survey confirmed the existence and extent of the hill.

The existence of this hill was particularly interesting since wells were located just to the south along the Sangamon River to pump groundwater from sand and gravel aquifers into the river to augment Decatur's surface-water supply during low flow. Several of these wells penetrate bedrock at elevations below 400 ft (122 m), and one penetrates below 350 ft (107 m), although the location and therefore the surface elevation of this well are now suspect. In any event a rather deep channel (±150 ft, or 46 m) separates the hill from the valley sides.

The other hill-channel (Kenney Valley) relation is that located mainly in northeastern Logan County. These features are documented (Hunt and Kempton, 1977), and the northwestward continuation of the Kenney Valley has been verified during this study. A smaller hill just southwest of Clinton in DeWitt County lies just to the southeast of the larger hill. The larger of these two hills rises just above an elevation of 500 ft (152 m), while the highest point penetrated on the smaller is 497 ft (151 m). Channels below elevation 400 ft (122 m) separate these two hills from the south valley wall. The main Mahomet Valley Lowland channel lies northeast and north of these hills.

There are a few wells that penetrate the Kenney Valley to the south; the lowest, about midway between the Mahomet and Mackinaw channels, encountered bedrock at an elevation of 377 ft (115 m). Along the main channel, in southwestern McLean County, a well about midway between the Clinton area and the Mackinaw Valley encountered bedrock at elevation 354 ft (108 m). There are relatively few wells in southwestern McLean County to adequately define the shape and depth of the main channel in that area. Therefore, although the existence of both channels is not questioned, there are too few data to assure that the configuration suggested by Figure 7 is necessarily accurate.

Tributary valleys. The tributary valleys that enter the Mahomet Valley Lowland can be characterized either as having large drainage basins, a generally dendritic pattern, and a graded floor, or as short and stubby with small headwaters and apparent steep gradients near the lowland margin. Characterization of some of the tributary valleys, particularly in the immediate vicinity of the lowland, is often difficult due to the lack and proper distribution of data at or near their confluence with the lowland. As with the main lowland, most tributary valleys tend to be oriented northeast-southwest or northwest-southeast; some are oriented only near the main lowland, others throughout their length.

The main lowland north and east of Rantoul gives the appearance of several tributary valleys converging to form the main lowland: the Danville from the southwest, the Onarga from the north-northeast, and the "main" valley from the east-northeast.

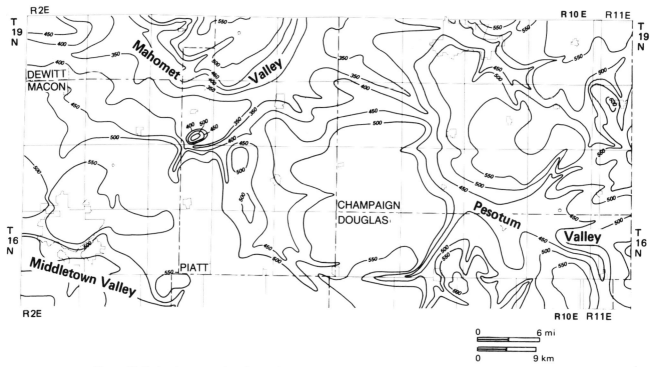

Figure 12. Bedrock topography of a portion of east-central Illinois, showing details of erosion remnant "hill" in Mahomet Valley, west-central Piatt County (from Kempton and others, 1982).

Because of poor data distribution, the exact shape or orientation of some tributaries is not clear, and the lowland side appears as a broad indentation (e.g., in southern Iroquois County and the area west of Rantoul). These may represent one or more tributaries with sharper valley sides than shown. None of these stubby or broader tributaries appears to have drainage basins of any significant size.

The only tributary south of Rantoul that has a well-developed drainage pattern is the Pesotum Valley (Figs. 7, and 12; Kempton and others, 1982). It tends to narrow just above (to the south of) its confluence with the Mahomet Valley Lowland. North of the Pesotum Valley, a small, narrow valley, oriented west-northwest, lies directly under Champaign-Urbana. This valley appears to have a rather gentle gradient, but is fairly short. The short, stubby tributary that enters the lowland just to the southwest of Monticello in the central Piatt County is the largest tributary valley mapped between Monticello and the Mackinaw Valley. It is this valley that Horberg connected with the Pesotum Valley (Fig. 1). Data along the west side of this tributary suggested a gentler, more irregular slope than the east side.

The Onarga Valley, the major northern tributary, appears to have its headwaters in Kankakee County. Preliminary remapping of the bedrock topography of that county by Kempton and Berg (1982) shows the headwaters valley to be a narrow, rather deep gorge cut into Silurian dolomite. The valley in Iroquois County and eastern Ford County is less well defined by available data, but does widen significantly toward its confluence with the Mahomet. Two reasonably well-defined tributaries, the Chatsworth and Kempton Valleys (Fig. 1), enter the Onarga from the northwest in northern Ford and western Iroquois Counties. Combined with these and smaller tributaries, the drainage basin of the Onarga Valley is the best developed and largest of the Mahomet tributary systems.

The confluence of the Danville Valley from the southeast with the Mahomet Lowland is the most complex of the tributaries. The lower reaches of the Danville Valley northwest of Danville, combined with a tributary from the southeast and a broader, less developed valley, from the east, form a broad lowland below an elevation of 450 ft (137 m). This area (below 450 ft, or 137 m) may be an extension of the broad bench that extends southwestward to south of Champaign. It contains what appears to be a complex of rather narrow channels below an elevation of 400 ft (122 m) that connect and enter the main Mahomet channel just southwest of Hoopeston.

The combination of the Onarga and Danville Valleys and their tributaries, together with the relatively narrow main Mahomet channel, gives the appearance of a headwaters region for the Mahomet Valley Lowland in Illinois.

Relation to Teays (Lafayette) and Ancient Mississippi (Mackinaw) bedrock valley systems

Horberg (1945) was the first to show a direct connection of the Teays system with the Ancient Mississippi system following the work of Fidlar (1943) and others in Indiana, and Stout and others (1943) and others in Ohio, that defined the Teays system in Ohio and traced it into and across central Indiana. Horberg confirmed the presence of a bedrock channel in the Lafayette area of west-central Indiana that connected it to the Mahomet Valley in Illinois (Horberg, 1945, Fig. 4), thereby inferring the existence of a major eastern bedrock valley system tributary to the Ancient Mississippi system (Horberg, 1945, Fig. 1), which had developed on the bedrock surface prior to Pleistocene glaciation of the region.

Horberg's map of the Teays Valley in central and west-central Indiana (Horberg 1945, Fig. 4) showed the valley to be between 5 and 10 mi (8 to 16 km) wide with relatively steep walls and a fairly flat bottom 3 to 5 mi (5 to 8 km) wide. The deepest part of the valley was shown to be below 350 ft (107 m) in elevation from about 10 mi (16 km) north of Lafayette to southeastern Benton County and then below 300 ft (91 m) across southern Benton County to the Indiana-Illinois state line.

Recent maps of the bedrock topography in Indiana—by Gray (1982; this volume) for the entire state, and by Bruns and others (1985) for the western part of the Teays Valley area—suggest somewhat different interpretations. Gray (1982) showed a rather complex, somewhat sinuous valley in southern Benton County with the deepest portion below 400 ft (122 m) at the state line and eastward to the Lafayette area, whereas Bruns and others (1985) showed a rather straight valley of generally uniform width straddling the Benton-Warren county line, with the deepest portion at the state line below 450 ft (137 m). Both maps, however, confirm a continuation of the valley westward from the Lafayette area; both show a broad, complex lowland in the Lafayette area narrowing westward into the Mahomet Valley segment. Evidence in Illinois suggests that the deepest portion in Illinois is below elevation 400 ft, although how far or continuous the deepest channel extends eastward at this point is speculative. Bruns and others (1985) did show areas below 400 ft (122 m) in the Lafayette, Indiana, area.

The continuation of the Mahomet Valley Lowland and the channel below elevation 350 ft (107 m) into the Mackinaw Valley portion of the Ancient Mississippi Bedrock Valley has not been questioned. However, the existence of the Kenney Valley and the sparsity of data defining the configuration of the "main" valley northwest of Clinton complicate evaluation of the relation of the two major bedrock valleys.

The confluence area forms a wide bedrock lowland emphasized by the Danvers Valley opening into the lowland from the northeast. All valleys appear graded to the Ancient Mississippi Bedrock Valley. However, the Ancient Mississippi is also complicated by the bedrock channel to the west of the Mackinaw channel that generally lies under the present course of the Illinois River (Horberg, 1950; Horberg and others, 1950), similar to the Mahomet-Kenney Valley relation.

Horberg (1945, 1950) does not show the 300-ft (91 m) contour extending north of the narrows of the Mackinaw Valley (Fig. 2). Recent mapping of the bedrock topography west of the Bloomington-Normal area (Richards and Visocky, 1982) shows no confirmed elevation below 353 ft (108 m) from the narrows northward. Therefore, both the Mahomet and Mackinaw appear to be similar in their general characteristics, although up valley the Mahomet may be slightly lower.

Interpretations and discussion

The morphology and development of the Mahomet Bedrock Valley were determined by a number of controlling factors and are the result of several major events that occurred over a long time interval; details of these factors and events are known only in a very general sense. The erosional cutting of the valley is obvious, but the number and timing of erosional events and their relation to specific erosional features in the valley and the deposits filling the valley are not known.

Preglacial drainage. Analysis of the revised bedrock topography suggests that a preglacial valley probably existed in the general area of the Mahomet Valley. The larger tributaries to the Mahomet have a normal dendritic drainage pattern, which in turn suggests that they were part of a larger drainage system. The drainage level of any such system is controversial and centers on the question of whether the deepest part of this and other similar bedrock valleys was entrenched prior to, as a result of, or after the first major midcontinent glaciation (see discussion in Willman and Frye, 1970). Evidence favors a relatively shallow entrenchment of the ancestral Mahomet drainage system (Frye, 1963; Willman and Frye, 1970), and we suggest that its headwaters probably existed south of the Kankakee Arch, a northwest-southeast–trending bedrock high in northeastern Illinois, and on the southwestern slope of the "Rensselaer Plateau" (Wayne, 1956), a broad bedrock upland in northwestern Indiana. The main valley may have continued eastward into Indiana along the later Mahomet Valley course, but more likely a headward tributary existed at that location.

The transverse character of the Mahomet Valley suggests the ancestral drainage system developed on relatively flat-lying younger Pennsylvanian rocks and through time was superposed across structures related to the La Salle Anticlinal Belt. While the northwest-southeast and southwest-northeast orientations of the Mahomet Valley Lowland and its tributaries suggest bedrock joint control, there are no specific data to support that. This stage of drainage development likely was ongoing from the late Tertiary well into the early Pleistocene.

Glacial disruption of drainage. Disruption of preglacial drainage probably was the result of early Pleistocene glaciation, but major stream piracies may have preceded such an event. Drainage disruption to the east (Gray, this volume; Bleuer, this volume) resulted in the diversion of regional meltwater drainage to the west, and meltwater from central and northern Indiana and Ohio drained into the Mahomet Valley from the "Teays" drainage system. The main course of the Mahomet Valley between Indiana and Illinois (Fig. 1) probably originated as the result of spillover across a low bedrock divide in western Indiana. These diversions apparently predate the glaciation that resulted in deposition of the West Lebanon till (Bleuer, this volume). Fullerton (1986) has inferred an earlier glaciation in Ohio. Thus, the Mahomet Bedrock Valley, as subsequently known, probably originated during this interval of earliest midcontinent glaciation.

Morphology of the Mahomet Valley. The morphology of the Mahomet Valley suggests a complex erosional history. A number of features within the Mahomet Valley Lowland are similar to described features associated with Wisconsinan meltwater drainageways, particularly those that carried large floods that catastrophically drained glacial lakes (Baker, 1978; Teller and Thorleifson, 1983; Kehew and Lord, 1986). Among these are the straight valley sides with short, steep (hanging?) tributaries, scoured bedrock benches, residual hills within the valley (many of which have a streamlined shape), secondary channels (anastomosing?) within the main valley, and the deep thalweg of the valley that appears to be a spillway-type channel cut in rock. This assemblage of erosional landforms indicates that large catastrophic floods probably are responsible for much of the major sculpting of Mahomet Valley Lowland. Such floods commonly are associated with sudden draining of large glacial lakes (Kehew and Lord, 1988); the features of the lowland are likely the result of catastrophic drainage of Lake Marion (Bleuer, this volume) and other early glacial lakes in Indiana and Ohio. The number and timing of these floods is not known, nor are we able to relate specific erosional features of the valley to the depositional fill in the valley.

Along the thalweg of the lower Illinois bedrock valley, where a detailed study has been made (Heigold and Ringler, 1979), there are large variations in declivity. In some localities the regional slope on the bedrock surface is removed. A possible explanation is that at one point in the history of the valley, a large volume of water of great velocity and short duration traveled this portion of the lowland. When the flood subsided, much of the load was dropped, covering and preserving bedrock features below. These features, therefore, are in harmony with the morphology of the Mahomet Valley Lowland. Additional, more detailed mapping of the bedrock surface within the Mahomet Valley Lowland may reveal similar features, including highs and lows in the thalweg.

In a seismic refraction study of the Meredosia Channel area of northwestern Illinois, McGinnis and Heigold (1974) found evidence of sizable grooves cut in Silurian dolomite bedrock, which they attributed to pre-Illinoian glacial erosion. There is no direct evidence for glacial scour of the present surface of the bedrock in the Mahomet Valley. The major features can be ascribed to erosion by water; the features described by McGinnis and Heigold in the Meredosia Channel also may be of catastrophic flood origin.

Isostatic movements of the Earth's crust in response to loading and subsequent unloading by Pleistocene glaciers must have had a great influence on Pleistocene drainage development and sediment accumulation. Bedrock gradients during Pleistocene time in many places were likely much different than they are today (McGinnis, 1968). It is therefore likely that a combination of forebulge development and erosion by torrential floods shaped the Mahomet Valley Lowland.

STRATIGRAPHY OF THE SEDIMENTARY (DRIFT) FILL

The development of the bedrock surface cannot be fully understood without considering the impact of glaciation and its resultant processes and deposits. The preceding discussion of the Mahomet Bedrock Lowland indicates that glaciation played a predominant role in shaping its existing configuration. Conversely, the distribution and preservation of Quaternary deposits has in part been controlled by the bedrock topography. Thus, working out this interrelation is the key to understanding the history of the Mahomet Valley and its fill.

The glacial deposits covering the bedrock surface in east-central Illinois range in thickness from a few feet to more than 400 ft (122 m) (Piskin and Bergstrom, 1975). Bedrock is exposed in a number of areas near the Mahomet Valley, particularly in Vermilion County, either as natural exposures along stream valleys or in strip mines, pits, and quarries. The thickest drift generally is found over the bedrock valleys, although the drift averages more than 150 ft thick throughout the region.

Horberg (1953) made the first comprehensive study of the subsurface deposits of northeastern Illinois, focusing much attention on the sequences observed in the Ancient Mississippi and Mahomet bedrock valleys. In this study, he defined the principal outwash in the Mahomet Valley, naming it the Mahomet Sand for the village of Mahomet in Champaign County. Previously, Horberg (1950, p. 51-52) had defined the Sankoty Sand in the Ancient Mississippi Valley. Horberg's criteria for recognizing multiple glaciations from subsurface data were: (1) weathered zones and their relative position, (2) organic and peat zones, (3) loess and fossiliferous silts, (4) stratigraphic succession, and (5) differences in physical properties. However, all glacial deposits were classified into chronostratigraphic units, based primarily on major weathered zones.

Quaternary deposits in Illinois are now subdivided and classified into lithostratigraphic, pedostratigraphic, chronostratigraphic, and morphostratigraphic units (Willman and Frye, 1970). The first two are the primary units utilized in this report (Fig. 13); they are placed in the chronostratigraphic classification to the extent possible. Formal units are based on work of Horberg (1953), Jacobs and Lineback (1969), Willman and Frye (1970), Johnson (1971), Johnson and others (1971), and Ford (1974). Subsequent work, particularly in the subsurface, suggests that the stratigraphic sequence is more complex, and some of the units are divisible into subunits (Johnson, 1986). These subunits are not firmly established, however, and generally cannot be related directly to the Mahomet Valley sequence and thus are not shown in Figure 13.

Regional stratigraphic framework

Maps and cross sections showing the sequence and distribution of these deposits are shown in Figures 14 through 20. Figure

9 shows the location of selected borings and wells shown in Figures 15 and 16 and the cross sections shown in Figures 17 through 19. Glacial tills and intercalated stratified deposits associated with the Mahomet Valley are included in the Banner, Glasford, and Wedron Formations (Fig. 13). The Wedron is in the Woodfordian Substage of the Wisconsinan Stage, the Glasford in the Illinoian Stage, and the Banner in an informal pre-Illinoian stage. The latter includes what earlier workers called Kansan and Nebraskan, but these chronostratigraphic units are not utilized because of stratigraphic and nomenclatural confusion in their type areas (Boellstorff, 1978; Hallberg, 1980, 1986) and because of problems of long-range correlation. Well-developed buried soils of regional significance and interglacial character, the Yarmouth and Sangamon, occur locally in the top of the Banner and Glasford Formations, respectively. Buried soils, either of less or uncertain significance in terms of degrees of development and climatic significance, occur at the top of the Smithboro and Vandalia Till Members (Fig. 13). Others may occur within the Banner Formation.

The Mahomet Sand Member (Fig. 14) is restricted to the Mahomet and Kenney bedrock valleys and here is extended to include a silt facies in tributary valleys and locally in the main valley. Its equivalent in the Mackinaw Valley, the Sankoty Sand Member, is restricted to the Mackinaw Valley and is, at least in part, correlative. The Hegeler and Belgium Members are restricted to an upland bedrock valley near Danville (Johnson,

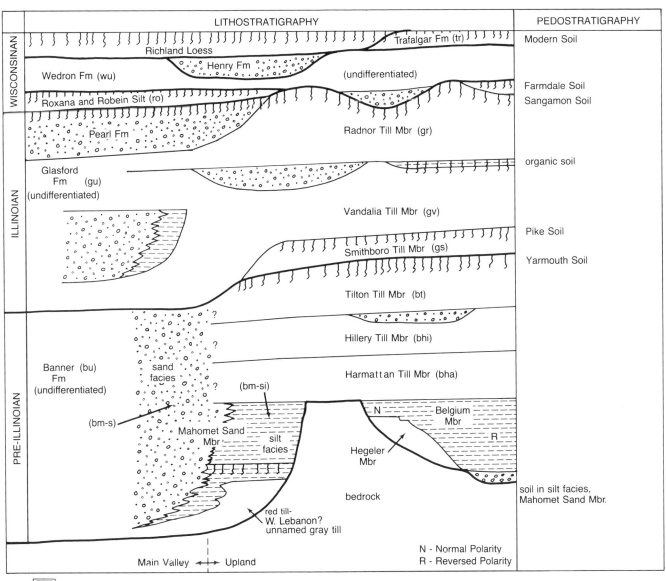

Figure 13. Pleistocene stratigraphy of east-central Illinois.

1971) and likely are correlative with parts of the Mahomet Sand Member. The Hegeler, originally interpreted to be till, now is considered to be the deposits of a large sediment gravity flow related to a valley margin, but derived from drift (Johnson, 1983). The Belgium Member (Fig. 13) consists of alluvial and ponded lake sediments that have reversed remnant magnetic polarity in one exposure and normal polarity in another. The change in polarity has been correlated tentatively with the Brunhes-Matuyama boundary at about 730,000 B.P. (Johnson, 1976).

Bleuer (1976, 1983, this volume) has described and informally named a reddish till in western Indiana the "West Lebanon." It occurs above proglacial silt with reversed polarity. Although not definitely identified in Illinois, the till may have been the parent material for the Hegeler sediment flow and, as discussed later, it may occur below the Mahomet Sand Member, as recorded in the well at Weldon (Fig. 16) and elsewhere.

The Banner Formation contains three formally defined till members in eastern Illinois: the Harmattan Till Member, a Lake Michigan lobe till, and the Hillery and Tilton Till Members, Huron-Erie lobe tills. Originally delineated and defined near and in now-abandoned strip mines near Danville (Johnson and others, 1971), these units have been extended regionally through subsurface studies (Kempton and others, 1980). The latter studies suggest a more complex stratigraphic framework, but because of limited subsurface control must be considered tentative and somewhat speculative (Johnson, 1986).

Deposits filling the Mahomet bedrock valley are known only in the subsurface. The major unit filling the valley is the Mahomet Sand Member (Horberg, 1953). The stratigraphic position of the Mahomet Sand is best known from tributary valleys where correlative slackwater sediments have been related to the regional stratigraphic sequence. The Mahomet Sand Member is described first, followed by a discussion of its stratigraphic position and relation to associated tills. Younger outwash sands, gravels, and tills of Illinoian and Wisconsinan age are considered in the context of the Mahomet Valley.

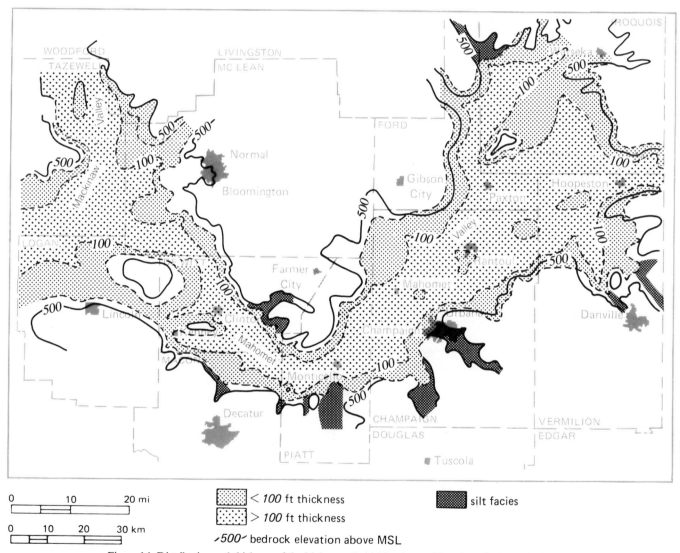

Figure 14. Distribution and thickness of the Mahomet Sand Member and location of related silt facies.

Banner Formation

Mahomet Sand Member.
The Mahomet Sand Member consists of two main facies: a sand facies that occurs in the main valley, and a silt facies that occurs in tributary valleys that were ponded during one or more sedimentation events and locally in the main valley.

Sand facies. The sand facies is restricted primarily to the main valley and is more than 100 ft (30 m) thick, except along the valley margins and where it overlies hills within the lowland (Fig. 14). In the deepest part of the valley, it commonly is more than 150 ft (45 m) thick. This facies is a coarse, gravelly sand that tends to fine upward. The fine- to medium-grained gravel component, as well as local coarse cobble gravels, are most conspicuous in the lower part of the unit but generally are present throughout. Sand and silt zones locally occur within the coarser deposits, but are always subordinate to sand or gravel in the main valley.

Manos (1961) studied the heavy-mineral composition of the sand facies to obtain information on the general provenance of the unit. His samples, taken from as deep in the valley as possible, contained a diverse heavy-mineral suite with abundant hornblende, garnet, and epidote. Manos observed that the suite was similar to that from the overlying tills, and concluded that the sand was derived from glacial drift. Subsequent observations have substantiated the glaciofluvial origin of the Mahomet Sand. As yet, nonglacial alluvial deposits have not been observed below the Mahomet Sand and above bedrock in any portion of the Mahomet Valley. Internal variations in clay mineral and carbonate composition have been recognized in the sand facies (H. D. Glass, personal communication), but sample control is not adequate to develop a regional mineral zonation. These mineralogic changes, as well as some abrupt textural changes, suggest that the sand facies is composed of multiple sedimentation units.

The lower portion of the sand facies appears continuous with sand in the Mahomet Valley in western Indiana (Fig. 15). The upper portion, however, does not appear to continue to the east, but instead appears to have been derived from a drainage source to the north and northeast in Iroquois County (Fig. 14) via the Onarga Valley and its tributaries. The uppermost surface of the Mahomet Sand attains its highest elevation within the main lowland, about 560 ft (171 m) in the vicinity of Hoopeston (Fig. 15) in northern Vermilion County (Kempton and others, 1981). This further supports a source just to the north and northeast. Down valley to the west-southwest, highest elevations range form 560 to 540 ft (171 to 166 m) west of Hoopeston, 550 to 530 ft (168 to 162 m) in the southeastern Ford and northeastern Champaign County (Paxton, Rantoul area), 540 to 510 ft (166 to 155 m) in central Champaign County (Champaign-Urbana, Mahomet area), 530 to 490 ft (162 to 149 m) in southwestern Champaign and central Piatt Counties (Monticello area), 510 to 470 ft (155 to 143 m) in DeWitt County (Hunt and Kempton, 1977), and from 500 ft (152 m) to less than 450 ft (137 m) at and westward from the junction with Mackinaw (Ancient Mississippi) Bedrock Valley. Sand and gravel, locally superimposed over the Mahomet (Sankoty) Sand in this junction area, up to elevation 550 ft (168 m) (Richard and Visocky, 1981), are likely to be within the Glasford Formation.

There is a suggestion that, in addition to the depositional irregularities of the surface of uppermost sand, additional local variations have been caused by subsequent differential compaction and/or erosion, both fluvial and glacial. In northern Vermilion County there is a suggestion of channel development on the surface (Kempton and others, 1981) near the center of the fill. Elsewhere, there may be channels along the margins of the fill (e.g., Fig. 17).

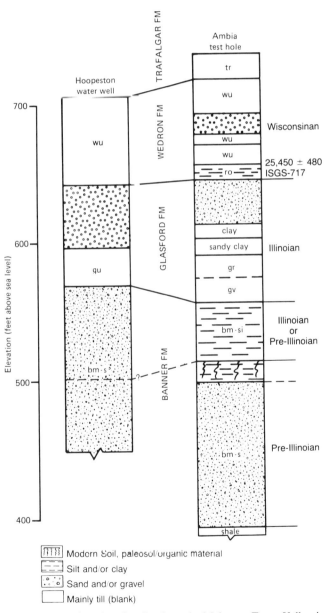

Figure 15. Stratigraphy of wells along the Mahomet-Teays Valley in Illinois and Indiana on either side of the state line.

Eastward, in western Indiana, the top of the Mahomet Sand or its equivalent is significantly lower (Fig. 15) and contains a cap of lacustrine silts and clays up to nearly the same elevation as the top of the Mahomet Sand in easternmost Illinois near Hoopeston. If the source of at least the uppermost portion of the Mahomet Sand in Illinois was from the north in Iroquois County (Illinois), this outwash and its associated discharge may have ponded the main Mahomet Valley east of Hoopeston in Indiana (Fig. 15) and eventually caused spillover from the Mahomet Valley through the Attica Cutoff to the Wabash Valley (Bleuer, this volume).

Silt facies. The silt facies consists of silt and clay that are present in some tributary valleys, particularly those on the south side of the main valley (Fig. 14). This facies generally is not as thick as the sand facies, but thicknesses greater than 40 ft (12 m) are known in several tributaries (e.g., Pesotum Valley). The sediment is generally calcareous, and cores show it is laminated silt and clay. In borings where clay mineral data are available, the data commonly indicate the facies is similar in composition to an overlying or underlying till.

Available data indicate the facies is primarily lacustrine in origin, and deposition took place in slackwater lakes when tributary valleys were ponded as a result of aggradation in the main valley or as a result of glacial ponding. A well-developed buried soil within the silt facies at Weldon (Fig. 16) suggests there were at least two episodes of sedimentation. The soil is about 10 ft (3 m) thick and is probably an interglacial alluvial soil.

As in the main valley, nonglacial alluvium is not know to occur below the silt facies in tributary valleys. Adequate sample

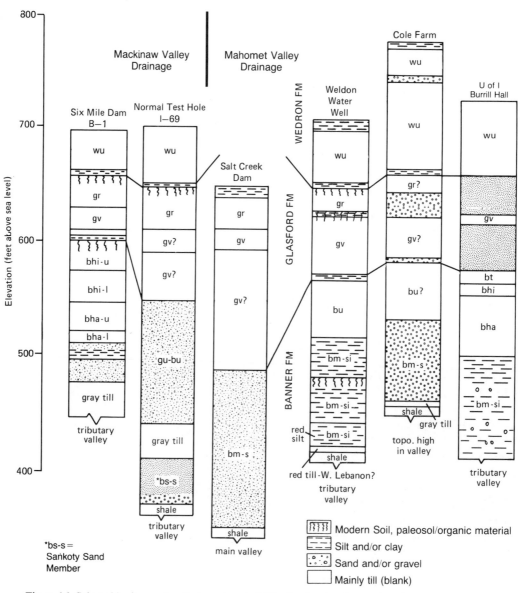

Figure 16. Selected borings and wells in east-central Illinois, showing the stratigraphic relations of the Mahomet Sand and interbedded and overlying deposits (see Fig. 9 for locations).

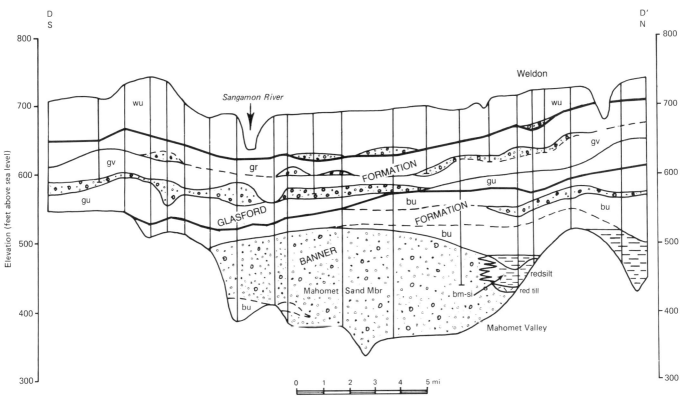

Figure 17. North-south cross section (D-D') across Mahomet Valley, southeastern DeWitt County to southwestern Piatt County (see Fig. 9 for location).

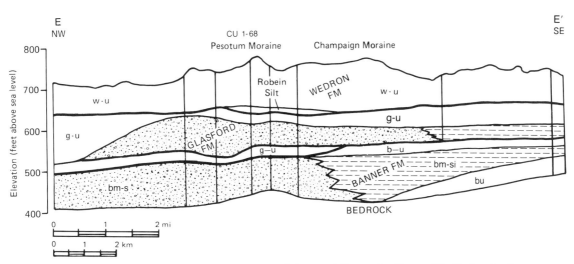

Figure 18. Northwest-southeast cross section (E-E') in central Champaign County (Fig. 9), showing the facies relation of sand and gravel in Mahomet Valley to silt and clay in tributary valley.

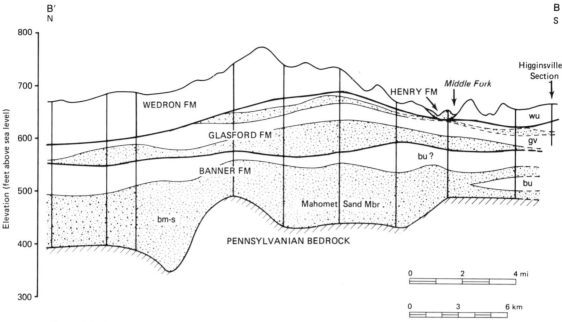

Figure 19. Cross section B-B' in northern Vermilion County (Fig. 9), showing a possible till wedge within Mahomet Sand at the south edge of the cross section.

control, however, is limited, and in many borings it is difficult to distinguish among weathered shale, nonglacial alluvium, and glacial slackwater lake deposits. The possibility that nonglacial alluvium locally may occur below the silt facies is suggested by deposits in the Harmattan strip mine at Danville (Johnson, 1971). The Belgium Member, primarily silt of lacustrine origin, overlies noncalcareous silt and conglomerate composed of local bedrock materials. The latter deposits are both alluvial and colluvial in origin. Although the Belgium Member and the silt facies of the Mahomet Member cannot as yet be directly correlated, they are similar in origin and physiographic position. This suggests that nonglacial alluvium may also occur below the silt facies of the Mahomet Member in some tributary valleys.

Sankoty Sand Member (Mackinaw bedrock valley). Recent work in the Mackinaw Valley (Richards and Visocky, 1982) suggest that the Sankoty Sand, as defined by Horberg (1950) and Horberg and others, (1950), is only the lower part of a complex of at least three outwash deposits. These deposits are locally separated by till but, where superimposed, may have a composite thickness of more than 150 ft (46 m). This entire thickness has been referred to as Sankoty, and correlation with the Mahomet Sand is implied (Willman and Frye, 1970). Where a lowermost sand (Fig. 16, Normal test hole) is overlain by till (below elevation 450 ft, or 137 m), it does appear distinctly different from the overlying sand and gravel. It is commonly pink or brown sand with some gravel. There is no obvious equivalent throughout the entire body of the Mahomet Sand. The middle outwash unit, which lies directly above the pink sand, is typical of most glacial outwash of the area and may be equivalent to a part of the Mahomet Sand (Fig. 16). The uppermost outwash unit, previously considered part of the Sankoty, may be correlative with the lowermost Glasford Formation outwash.

Ongoing stratigraphic studies of the confluence area of the Mahomet and Mackinaw Bedrock Valleys by the Illinois State Geological Survey may help to resolve the relation of the Mahomet Sand to the various units of the Sankoty Sand. However, existing data from borings in the Mackinaw Bedrock Valley drainage to the north of the confluence area (Fig. 16) are helpful in understanding the regional stratigraphy. The principal complications in relating the stratigraphic sequences in the Mackinaw drainage to the Mahomet may be the influence of northwestern source deposits in the Mackinaw drainage on the northeastern source deposits in the Mahomet.

Till stratigraphy. Regional relations. The stratigraphic position of the Mahomet Sand Member is crucial with regard to the age of the Mahomet bedrock valley, and to relations among glacial, glaciofluvial, and fluvial events. Horberg (1953) considered the Mahomet Sand to be older than any till in Illinois, although he realized it might have been derived from a glacial source to the northeast. He considered till that overlies the sand in a few borings to be Nebraskan. Most recent workers have considered the Mahomet, as well as the Sankoty Sand, to be pro-Kansan outwash (Willman and Frye, 1970). Our deep stratigraphic control is extremely limited, and for this reason we have included two borings in the Mackinaw Valley near its confluence with the Mahomet Valley (Figs. 9, 16). These borings contain Sankoty Sand rather than Mahomet Sand. Available data suggest that both the Mahomet and Sankoty Sand Members in places overlie one or more unnamed tills, contain interbedded till possibly related to a Lake Michigan Lobe event, and are overlain by the Harmattan Till Member, a younger Banner Formation till or the Glasford Formation (Fig. 16).

Tills below or interbedded with Mahomet and Sankoty Sand Members. In several borings, till or till-like material occurs below or is interbedded with the Mahomet and Sankoty Members. Although identified as till in Figures 13, 16, and 17, the samples are difficult to interpret, and some of the materials may not actually be till. All are drift-derived, however, and thus are related to or postdate one or more early glacial events. All of these occurrences are from borings in marginal areas of the main valley or its tributary valleys, or in the Mackinaw Valley near its confluence with the Mahomet.

A boring for Six-Mile Dam (Fig. 9), located in a tributary valley north of the confluence of the Mahomet Valley and the Mackinaw Valley, terminated in 30 ft of gray till (Fig. 16). The till is a pebbly loam and contains a moderate amount of expandable clay minerals. It is similar in character to till that is correlated with the Harmattan Till Member and occurs higher in the boring above relatively thin sand and silt correlated with the Sankoty Member. The lower till is thought to have a northern source and probably was deposited by the Lake Michigan Lobe.

A test boring in the Normal well field, 15 mi (24 km) to the southwest and closer to the Mahomet Valley but still in the Mackinaw Valley (Fig. 9), contains a gray till interbedded between the lowermost and middle units of the Sankoty Sand (Fig. 16). The till contains large amounts of expandable clay minerals and is most similar to till described by Johnson (1964) some 50 mi (80 km) to the southwest. Based on heavy-mineral studies, Johnson (1976) related the latter till to a northern (Lake Michigan Lobe) source and later correlated it with the Harmattan. Another alternative is that these older gray tills in central Illinois may be lateral facies equivalents of the reddish West Lebanon till of Indiana. The portion of the ice sheet flowing west of central Michigan would not have entrained reddish sediment from the

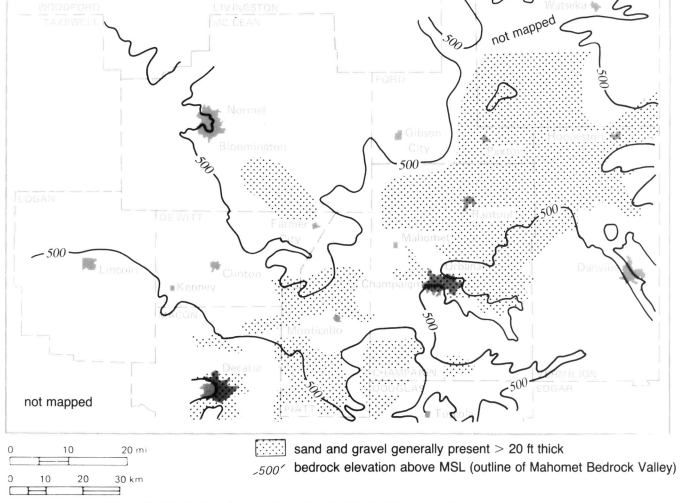

Figure 20. Distribution of sand and gravel within Glasford Formation (Illinoian) in relation to the Mahomet Bedrock Valley.

central portion of the Michigan Basin. It is not known, however, whether any of these tills are older than the Brunhes-Matuyama boundary as is the West Lebanon (Bleuer, this volume).

Farther east, a water well for the village of Weldon (Fig. 9) contains a thick sequence of silt facies of the Mahomet Member that overlies a thin (6 ft, or 2 m) red till over shale (Fig. 16). The basal 20 ft (6 m) of the silt facies is also reddish and probably is related to the underlying till. Although little is known about the composition of the till, its stratigraphic position, color, and occurrence near the base of the bedrock valley suggest a possible correlation with the West Lebanon till of western Indiana. Borings across the Mahomet Valley to the south also suggest that till occurs below the Mahomet Sand (Fig. 17), but little is known with regard to the character of the material.

Another boring where till occurs below the Mahomet Sand is in Champaign County, the Cole Farm well (Fig. 16). This well is located over a remnant bedrock hill within the broad Mahomet Valley Lowland. Sampling and coring in the well encountered a thin (3 ft, or 1 m) gray diamicton above shale. The material looks like till, contains igneous pebbles, and has a relatively small carbonate content. The clay mineral composition, however, reflects the local shale bedrock (H. D. Glass, personal communication). A well in a tributary valley at Champaign-Urbana appears to contain till below the silt facies of the Mahomet Member (Fig. 18).

Work in northern Vermilion County suggests that till is interbedded within the Mahomet Sand Member (Fig. 19; Kempton and others, 1981). Sample control is poor, and little is known with regard to the composition, character, and age of these materials.

Tills overlying the Mahomet Sand Member. The Mahomet Sand Member is overlain by till of the Banner Formation and/or the Glasford Formation. Where sample control is poor, particularly over the main valley, it is not possible to identify which till member occurs above the contact; in some cases even the formational identification is uncertain. In those borings with the best stratigraphic control, the Harmattan Till Member is the oldest unit identified above the Mahomet or Sankoty Members. This situation occurs at Six-Mile Dam (Fig. 16), where two units of Harmattan overlie a sand equivalent to a part of the Sankoty Sand Member; these units, in turn, are overlain by two units of the Hillery Till Member. In the well for Burrill Hall at the University of Illinois (Fig. 16), the Harmattan, Hillery, and Tilton Till Members all overlie the silt facies of the Mahomet Member. However, both of these borings are in tributary valleys, and the stratigraphic situation above the sand facies in the main valley is less certain.

Early studies (Horberg, 1953; Stephenson, 1967), as well as more recent studies (Hunt and Kempton, 1977; Kempton and others, 1982), have suggested that one or more pre-Illinoian Banner Formation tills that are well documented adjacent to the Mahomet Valley cross over the valley and bury the Mahomet Sand, as suggested in Figures 17 and 19. Stratigraphic interpretation of the till above the Mahomet Sand in these studies is not firm, however, and the capping till may be younger. In borings where sample control is better (Normal Test Hole and Salt Creek Dam, Fig. 16; Ambia Test Hole, Fig. 15; boring CU 1-68 in Fig. 18), the till immediately above the Mahomet Sand is Glasford Formation (Illinoian) based on available composition data. Thus, it is possible that the Banner Formation tills that occur over the silt facies in tributaries of the Mahomet Valley (Six-Mile Dam, Burrill Hall, Fig. 16) were cut out of the main valley during and following one or more pre-Illinoian glaciations. If so, the situation would have been similar to major meltwater valleys of the Wisconsinan such as the Illinois and Wabash Valleys; i.e., Wisconsinan tills occur in their valley walls but are not part of their thick valley fills.

It appears that a major part of the Mahomet Sand was derived from both Lake Michigan Lobe source tills, the Harmattan and West Lebanon–equivalent and Huron-Erie Lobe tills, the Hillery and Tilton tills in Illinois, and the Brookston (Bleuer, this volume) in Indiana. The sand and gravel fill probably represents valley-train deposition during several episodes of pre-Illinoian ice margin advance and retreat. As discussed below, Illinoian outwash may be included. Better subsurface data will be necessary before the stratigraphic interpretations and their implications can be more firmly established.

Glasford Formation

The central and eastern parts of the Mahomet Valley contain significant deposits of Illinoian sand and gravel (Fig. 18) that are part of the Glasford Formation, i.e., they occur above or between Illinoian tills (Fig. 13). The western portion of the valley, however, contains little or no sand and gravel in the Glasford Formation, (e.g., the Salt Creek Dam, Normal Test Hole, and Six-Mile Dam borings, Fig. 16). This part of the valley was filled with thick Glasford Formation tills and did not function as a drainageway during the Illinoian, except possibly during the earliest phase of Illinoian glaciation. In areas where Glasford Formation till directly overlies the Mahomet Sand in the main valley, the upper part of the Mahomet Sand likely is proglacial (Illinoian) outwash that was eventually overridden by the Illinoian ice sheet.

Sand and gravel bodies in the central and eastern portions of the valley have a northeast-southwest orientation (Fig. 20), are laterally continuous over much of the area (the map shows only deposits thicker than 20 ft, or 6 m), and are not confined to the Mahomet Valley Lowland. The sand and gravel is best developed in the eastern part of the valley (Figs. 15, 16, 19). No stratigraphic distinction of sand and gravel within the Glasford Formation is made in Figure 20, but in most cases the deposits are associated with the Vandalia Till Member, either at its base, within, or above the till (Fig. 16, Cole Farm, University of Illinois Burrill Hall). It is not clear whether any sand and gravel is related to an earlier Illinoian advance that resulted in deposition of Smithboro Till. Some is related to the younger Radnor Till, which is the youngest Illinoian till recognized in this part of Illinois, and at the Indiana State line (Fig. 15), sand and gravel occurs above the

Radnor Till Member. Better subsurface control is required to evaluate the internal stratigraphic position of the Glasford Formation outwash, and such data would be significant for shallow groundwater exploration.

The northeast-southwest–trending outwash bodies appear to be related to, and some are undoubtedly a northward extension of, sand and gravel (Pearl Formation; Killey and Lineback, 1983) associated with the "ridged drift" of the Illinoian drift plain in southern Illinois (Fig. 21). The latter deposits are a series of ice-contact sands, gravels, and silts that are closely related to the Vandalia Till, and that accumulated during deglaciation and stagnation of the Illinoian ice sheet (Leighton, 1959; Leighton and Brophy, 1961; Jacobs and Lineback, 1969). Thus, based on this association, most of the sand and gravel in the Glasford Formation within the Mahomet Valley Lowland is considered to be outwash related to Illinoian glaciation. Some of it, however, may have accumulated during the Sangamonian and be nonglacial in origin.

North of the "ridged drift," a broad lowland probably existed more or less over the Mahomet bedrock valley during the Illinoian. Meltwaters accumulated in valleys within the lowland and in subglacial and ice-walled channels that drained to the southwest. The main source of outwash appears to have been from the north in Iroquois County (Fig. 20). Northward tracing

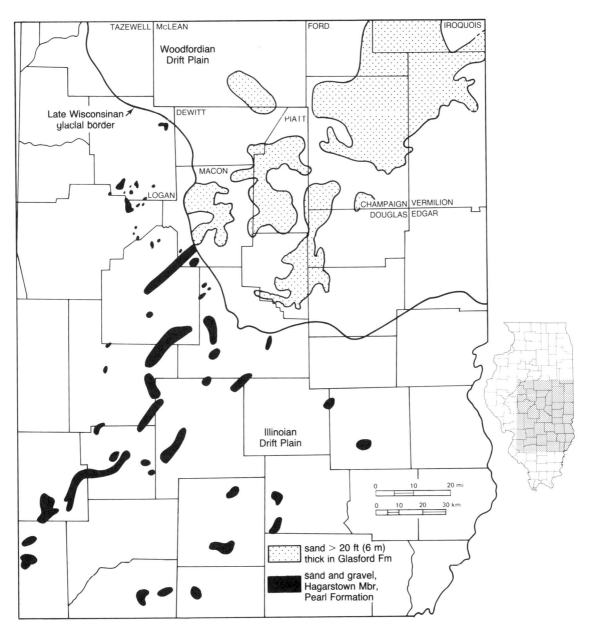

Figure 21. Subsurface Glasford Formation sand in relation to sand and gravel in the "ridged drift" area of south-central Illinois (Hagerstown Member of Pearl Formation; after Killey and Lineback, 1983).

of these deposits may supply further understanding of their origins.

Sub-Wedron Formation surface

The sub-Wedron surface (Fig. 21), as defined by the occurrence of Farmdale Soil in Robein or Roxana Silt or Sangamon Soil in Illinoian deposits (Fig. 13), is a composite surface that reflects the constructional topography that formed as a result of Illinoian glaciation, and modification of that surface during the Sangamonian and early and middle Wisconsinan times. It occurs at an elevation of approximately 650 ft (198 m) over a broad area (Fig. 22). The surface crosses the Mahomet Valley with little or no deviation in slope (Horberg, 1945, 1953), confirming that the valley was filled completely by the end of the Illinoian. Bodies of sand and gravel in the Wedron or Henry Formations that are related to Woodfordian (late Wisconsinan) glaciation (Fig. 13) bear no direct relation to the Mahomet Bedrock Valley.

Summary and discussion

Preglacial deposits. Assuming a preglacial valley existed, alluvial deposits would have accumulated in it. Such deposits are not known. The lack of preglacial alluvium in the main valley, however, is not surprising, given the subsequent meltwater floods that drained through it, and the likelihood that such alluvium would have been eroded. Preservation potential of preglacial alluvium is much greater in tributary valleys, particularly any that were ponded and became the site of slackwater lake sedimentation. Available data have not confirmed the existence of such deposits in tributary valleys, except in the strip mine at Danville, and the latter deposits have not been tied directly to the Mahomet

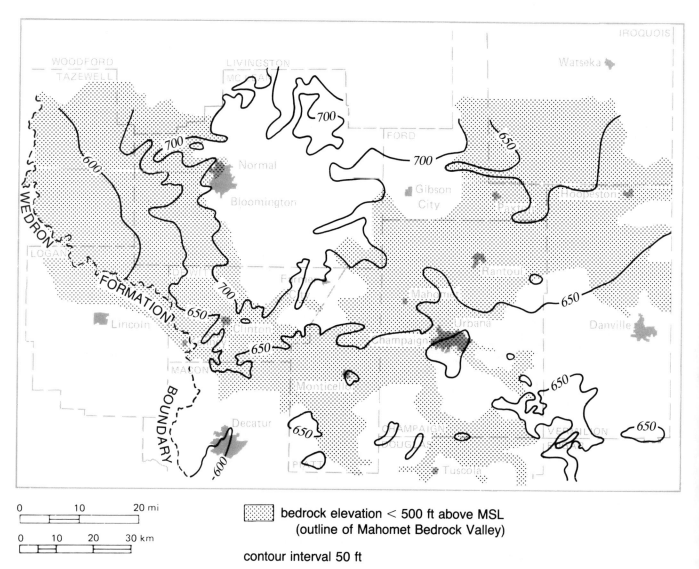

Figure 22. Elevation of base of Wedron Formation (late Wisconsinan), showing lack of relation to the Mahomet Valley.

Valley. Our data base is limited, however, and controlled drilling and sampling are needed to evaluate the presence or absence of preglacial alluvium below the glacial sequence in tributary valleys. If present, it would be possible to determine the general drainage level in the ancestral Mahomet Valley prior to major modification as a result of glaciation.

Glacial deposits. Available data suggest the basal part of the Mahomet Sand Member, the Banner Formation, overlies and likely is associated with one or more tills, possibly the West Lebanon till and a correlative gray till of Lake Michigan Lobe source. It also contains interbedded till of probable Lake Michigan Lobe source that may correlate with the older gray till just referred to or the younger Harmattan Till Member. The main part of the unit is likely associated with the Harmattan Till Member, also a Lake Michigan Lobe till, as well as the easternsource pre-Illinoian tills, the Hillery and Tilton Tills in Illinois, and the Brookston till of Indiana. The upper part of the Mahomet Member may be Illinoian, if the till overlying the unit is Illinoian and not pre-Illinoian, as suggested in prior studies. The Mahomet Sand was probably deposited during at least two major pre-Illinoian glaciations, separated by an interglacial episode, and likely one or more additional major glaciations.

The Mahomet Sand fill for the most part is interpreted to be outwash deposited by braided glacial rivers or locally by prograding deltas (late phase). Valley-train aggradation in the valley was associated with several pre-Illinoian glaciations and probably the earliest Illinoian glaciation. Slackwater lakes formed in tributary valleys during at least two of these events. Erosional unconformities must be present within the sequence but they cannot be documented with our available data base. Most till deposited in the valley was removed during times of deglaciation, or subsequent interglaciation.

In addition to Illinoian sand and gravel that may be included in the Mahomet Sand, sand and gravel deposits in the central and eastern portion of the Mahomet Valley are clearly part of the Illinoian Glasford Formation, inasmuch as they are underlain by Illinoian till. These deposits appear to be genetically related to the "ridged drift" of south-central Illinois and probably were deposited in a system of subglacial, englacial, and supraglacial channels related to the Vandalia and Radnor tills. The Mahomet Valley was completely buried and filled with drift after Illinoian glaciation.

HISTORY OF THE MAHOMET VALLEY

The history of the Mahomet bedrock valley is complex; many events are poorly recorded, and some are probably not recorded at all. In addition, the sedimentary record is not easily accessible for study. As a result, aspects of its history cannot be determined and much must be inferred. The following discussion summarizes what is known and emphasizes the constraints that can be placed on the history, given our current data base.

Early development—The preglacial drainage system

Although little evidence remains, it is inferred that an ancestral preglacial drainage system developed in east-central Illinois; it included a major tributary approximately in the same area of the Mahomet Bedrock Valley. The margins of this ancestral Mahomet drainage system likely were located in northeastern Illinois, western Indiana, and south-central Illinois, where a bedrock divide separated northward and eastward drainage from westward drainage into the ancestral Mahomet system. The system developed on a cover of Pennsylvanian rocks and later was superposed across older rocks related to the La Salle Anticlinal Belt. The level of this system probably was relatively high compared to the current deep Bedrock Valley, and drainage development likely was ongoing from late Tertiary well into the early Pleistocene.

Later development—Erosional valley modification

Modification of the ancestral Mahomet drainage system started prior to glaciation of the local area, but occurred in response to one or more early eastern midcontinent glaciations that are poorly known and dated. Such events disrupted preglacial drainage systems and eventually diverted northward drainage in Ohio to the west (Goldthwait, this volume). Spillover from the Battleground Lowland in Indiana (Bleuer, this volume) to the Mahomet Bedrock Valley occurred during one of these early glaciations. This established the Marion-Mahomet or "Teays" drainage system of Bleuer (this volume) and initiated major erosional modification of the preglacial Mahomet drainage system. These events apparently are older than the West Lebanon till (Bleuer, this volume), and thus must have taken place prior to 730,000 B.P.

The Mahomet Bedrock Valley, as subsequently known, probably originated during this interval. The greatly increased discharge and complications associated with isostatic adjustments of the land surface (Frye, 1963; McGinnis, 1968) resulted in major deepening and widening of the Mahomet Valley. The valley must have been deepened to at least an elevation near 400 ft (122 m), if the materials below the Mahomet Sand Member near Weldon are till (Figs. 16, 17). The varied nature of the topography suggests several episodes of cutting and filling, and the size and morphologic features of the valley suggest floods of extremely large discharge. The latter probably resulted from catastrophic drainage of Lake Marion (Bleuer, this volume) and other early glacial lakes in Indiana and Ohio.

Latest development—aggradational valley fill

Deposition of Mahomet Sand probably began during these events and likely was involved in the cut-and-fill history. Assuming the red till and slackwater lake deposits at Weldon (Fig. 16) are related to the West Lebanon event, deposition must have begun prior to 730,000 B.P. (Bleuer, 1976, 1983). Gray slackwater silts occur above the red silts at Weldon and below a major buried soil. Gray till also occurs near the base of the valley. These latter deposits probably resulted from an early event of the Lake Michigan Lobe and may be closely related in time to the West Lebanon glaciation. Bleuer (1983) wrote that the red coloration of the West Lebanon was derived from Jurassic red beds in the Michigan Basin in central Michigan. The lower gray tills in Illinois may be the same age, but have a composition and color that reflect Paleozoic bedrock to the west and southwest of the Michigan Basin. These early events were followed by at least one major interglacial period, as suggested by the buried soil at Weldon (Fig. 16), and by a site in western Indiana where a soil that developed in alluvium and colluvium occurs above West Lebanon till and stratigraphically below Harmattan Till (Bleuer, 1983; Follmer, 1983).

Later valley-train deposition of Mahomet Sand appears closely related to subsequent glacial events that resulted in deposition of the Harmattan, Hillery, Tilton, and Brookston Tills. The internal stratigraphy and history of the Harmattan is uncertain, but it may represent more than one event (Johnson, 1986). The interbedded till at Normal and the relatively high-level till at Cole Farm (Fig. 16) might relate to an early Harmattan event. At a later time, outwash of proglacial origin was overridden by the Harmattan ice sheet, and Harmattan till was deposited over and south of the valley in Illinois. This is based on the interpretations that the oldest till overlying the Mahomet Sand Member in the main valley is pre-Illinoian and part of the Banner Formation (Figs. 17, 19). If correct, this event then appears to have terminated the Mahomet Valley's role as a major meltwater drainageway. If these tills are younger (Illinoian), the Mahomet Valley served as a meltwater drainageway during subsequent pre-Illinoian glaciations and deglaciation and early Illinoian glaciation, as well as a major alluvial valley during the Yarmouthian.

During the Illinoian, the Mahomet Valley was overridden and the western portion was filled with till. Outwash sedimentation, however, continued over the eastern and central portions of the Mahomet Valley (Fig. 20). Meltwater drainage from the Lake Michigan Lobe was concentrated in a broad lowland or sag that existed over this portion of the valley. In addition to till, sand and gravel were deposited in a complex system of supraglacial, englacial, and subglacial channels, many of which probably were genetically related to deposition of the Hagarstown Member of the Pearl Formation in the "ridged drift" area of south-central Illinois (Fig. 21).

The Mahomet Bedrock Valley had no apparent influence on drainage during the Wisconsinan Age. Outwash and alluvial deposits are located in positions that are the result of Illinoian deglaciation and younger Sangamonian and Wisconsinan events. The Ancient Mississippi Bedrock Valley, however, probably continued to control drainage from the north and west until finally diverted to its present course by Woodfordian glaciation (Glass and others, 1964).

HYDROGEOLOGY

The continuing development of the details of the bedrock topography, stratigraphic framework of the contained sediments, and the geologic history of the Mahomet Valley Lowland goes hand-in-hand with developing a better understanding of groundwater movement and availability within the lowland deposits. The coarse outwash deposits that compose a significant portion of the valley fill of the lowland and its "trunk" valley to the west (Mackinaw) are the most extensive, highly productive sand and gravel aquifers in the southern three-fourths of Illinois, and possibly within the entire state.

The following discussion provides a general summary of the hydrogeologic setting based on the geologic framework presented above and on available hydrologic data.

Summary of aquifer characteristics

Mahomet Sand (Banner Formation). The Mahomet Sand is the most important aquifer in east-central Illinois. It fills the deeper parts of the Mahomet Valley Lowland throughout its length and in the tributary Onarga Valley and the Kenney Valley (Fig. 14). Its thickness and areal extent are controlled by the elevations of its upper surface and the configuration and elevation of the bedrock surface. Therefore, the Mahomet Sand is locally as much as 200 ft (61 m) thick and averages close to 100 ft (30 m) thick. In areas of bedrock benches and hills within the valley, near the lowland rim, and where the surface of the Mahomet Sand has been eroded, the thickness is less.

The Mahomet Sand is composed primarily of clean sand and gravel with only minor amounts of fines. Overall, its composition tends to fine down valley (east to west) and upward, with sand often predominant in the upper 50 ft (15 m) along most of the valley and predominant throughout the entire sequence generally northwest of Clinton in the main channel. The finer upper portion ranges up to 50 ft (15 m) thick, although only the upper 10 to 25 ft (3 to 8 m) may be silty. Therefore, along the margins of the deposit where it measures less than 50 ft (15 m) thick, the water-yielding potential may be greatly reduced. Also, the overall productivity appears to be greater in the valley segment from Hoopeston westward to near Clinton than the down-valley (northwestward) segment from Clinton to the Mackinaw Valley junction. Since most of the tributary valleys contain predominantly ponded, fine-textured, water-laid sediments stratigraphically related to the Mahomet Sand, their potential productivity is low.

Glasford Formation aquifers. Whereas persistent sand and gravel deposits have been identified in two or more stratigraphic positions within the Glasford Formation, the most widespread and generally thickest of these is associated with the Vandalia Till Member at or near the base of the Glasford (Fig. 13). Thinner, intra-Glasford sand and gravel deposits are locally present in the Glasford Formation between the Vandalia and Radnor Till Members and at the top of the formation.

The sand and gravel related to the Vandalia Till Member appears to be the most extensive aquifer (Fig. 20), aside form the Mahomet Sand, over the Mahomet Lowland. Although its distribution does not appear to have been controlled directly by the lowland configuration (Fig. 21), it is coarsest and thickest (about 100 ft, or 30 m) just west of Champaign-Urbana (Fig. 18), where it is locally separated from the Mahomet Sand by only a few feet of till. Throughout the area shown in Figure 20, it commonly ranges in thickness from 5 ft (1.5 m) to locally more than 60 ft (18 m).

The outwash related to the Vandalia Till, while normally composed of sand and fine to coarse gravel, is highly variable, particularly around the margins of the deposits. In addition it is more variable in its thickness and distribution than the Mahomet Sand. If portions of the outwash are associated with the "ridged drift" deposits farther south, with an ice-contact origin as described previously, this may explain the variability. Although the outwash deposits associated with the Vandalia are locally important aquifers, particularly where they are really extensive, overall they are of much less significance than the Mahomet Sand.

Evaluation of groundwater resources

There have been considerable amounts of data generated on the hydrologic properties of the Mahomet Sand and the aquifers of the Glasford Formation. Much of this data has been obtained from routine production tests of wells. The hydrologic information summarized here has largely been taken from Visocky and Schicht, (1969), Gibb (1970), Sanderson (1971), Woller (1975), and Kempton and others (1982). Data presented in Figure 23 include some updated values provided by A. P. Visocky (personal communication).

Mahomet Sand. Figure 23 shows the location of the principal municipal and industrial groundwater supplies developed in the Mahomet Valley, along with representative values for transmissivity and pumpage as of 1982. It should be noted that hundreds of farm and domestic wells and numerous wells for small communities and industries (including irrigation) are scattered throughout the lowlands.

Total groundwater pumpage from the Mahomet Sand was estimated to be about 42 mgd (million gallons per day) (2×10^8 liters per day [l/d]) in 1982. Maximum pumpage at that time for Champaign-Urbana (the largest single user) was in excess of 17 mgd (64×10^6 l/d), of which 2 mgd (7×10^6 l/d) is from the Glasford Formation aquifer. Recent peak pumpage has reached nearly 30 mgd (11×10^7 l/d) at Champaign-Urbana; therefore, it is likely that total pumpage from the Mahomet Sand throughout its extent now approaches 80 mgd (30×10^7 l/d). Maximum pumpage may be as much as 3,000 gpm (gallons per minute) (1×10^4 l/m) from individual wells.

Pumping test data, mainly from the higher capacity wells in the eastern half of Mahomet Valley, show transmissivities of as much as 51×10^4 gpd/ft (gallons per day per foot) ($5 \times 10^{-2} m^2/s$), hydraulic conductivities of as much as 4,780 gpd/ft (2×10^{-3} m/s), and storage coefficients ranging from 2×10^{-5} to 2×10^{-3}, with specific capacities as much as 207 gpm/ft (786 lpm/m) reported. The median for hydraulic conductivity is 2,920 gpd/ft^2 (1.3 m/s) for the Mahomet Sand in the area from Monticello to Champaign-Urbana (Kempton and others, 1982). The reported range of transmissivities is from 4,860 to 550,000 gpd/ft (7×10^{-4} to 8×10^{-2} m/s).

Water levels in the Mahomet Sand range from elevations of 700 to 670 ft (213 to 204 m) in the Hoopston-Paxton area on the east, 690 to 660 ft (210 to 201 m) in the vicinity of Rantoul, 670 to 640 ft (204 to 195 m) in the Champaign-Urbana and Mahomet area, 650 to 620 ft (198 to 189 m) in the Monticello area, 630 to 595 ft (192 to 181 m) in DeWitt County (Clinton area), and between 600 to 550 ft (183 to 168 m) in the junction area with the Mackinaw Bedrock Valley in southwestern McLean and southern Tazewell Counties. Hydrostatic level above the sand is about 140 ft (43 m) in the Hoopston-Paxton area to 100 ft (30 m) in the vicinity of the Mahomet-Mackinaw Bedrock Valley junction. The potentiometric surface, coincidentally, therefore roughly parallels the surface of the Mahomet Sand, with a gradient of slightly greater than 1 ft/mi (19 cm/km).

Glasford Formation aquifers. Numerous small communities and industries and farm and domestic wells obtain their water supplies from the aquifers within the Glasford Formation. The most important of these aquifers is that at or near the base of the Vandalia Till (Fig. 20). The initial municipal well field for Champaign-Urbana was developed in this aquifer; approximately 2 mgd (8×10^6 l/d) were pumped in 1982. The village of Hoopeston also utilizes this aquifer; peak pumpage is about 2 mgd (8×10^6 l/d). Outside the Mahomet Valley Lowland, it is the most important aquifer in east-central Illinois.

Pumping test data show transmissivities as much as 233,000 gpd/ft (3×10^{-2} m/s), although frequently the values are much lower. Hydraulic conductivities in the Glasford aquifer throughout east-central Illinois range to as much as 4,660 gpd/ft^2 (2×10^{-3} m/s) and a median of 885 gpd/ft^2 (4×10^{-4} m/s), with storage coefficients ranging from 1×10^{-5} to 8×10^{-2}. Specific capacities range from 0.4 to 110 gpm/ft (5 to 1×10^3 lpm/m). The reported range of transmissivities is from 694 to 233,000 gpd/ft (1×10^{-4} to 3×10^{-2} m/s).

The water-yielding properties of the principal Glasford aquifer are much more variable than those of the Mahomet Sand. However, depending on the extent and thickness of specific areas of coarse-textured deposits, such as just northwest of Champaign-Urbana, it may be highly productive; it can also be marginal for municipal development.

Water levels in the main Glasford Formation aquifer are almost always 5 to 30 ft (1.5 to 9 m) above those of the Mahomet Sand. Since the surface of the Glasford aquifer is generally between elevations 650 and 590 ft (198 and 180 m), its surface is 50 to 100 ft (15 to 30 m) above the top of the Mahomet Sand. The potentiometric surface of the Glasford aquifer ranges from about

40 to 80 ft (12 to 24 m) above the top of the aquifer, or between elevations 710 ft (216 m) in northern Vermilion County and 660 ft (201 m) locally in areas south of the Mahomet Valley. While water levels in the Glasford are normally higher than those in the Mahomet Sand, the water levels in the Mahomet Sand approach those in the Glasford in northern Vermilion County and in the Champaign-Urbana area. The till separating the two aquifers is locally only a few feet thick in the Champaign-Urbana area, and local pumpage has probably decreased hydrostatic head differences.

Recharge and groundwater movement. Recharge to buried aquifers in the Mahomet Valley Lowland occurs as vertical leakage of water from precipitation through the overlying deposits into the aquifers. Of the water that enters these deposits, an average annual rate of 107,000 gpd/mi^2 (250,000 lpd/km^2) has been calculated as entering the Mahomet Sand in east-central Illinois (Visocky and Schicht, 1969). Recently determined coefficients of vertical hydraulic conductivity of the confining beds range from 2.12×10^{-3} to 0.4 gpd/ft^2 (1×10^{-9} to 2×10^{-7} m/s). These values yield recharge rates on the same order of magnitude as the earlier calculations based on water balances. Due to the similarity of overlying deposits, the leakage rate of 107,000 gpd/mi^2 (25,000 lpd/km^2) is assumed to hold for the entire Mahomet Valley Bedrock Lowland.

Groundwater movement in the Mahomet Sand appears to be semi-isolated from the shallower local flow systems generally

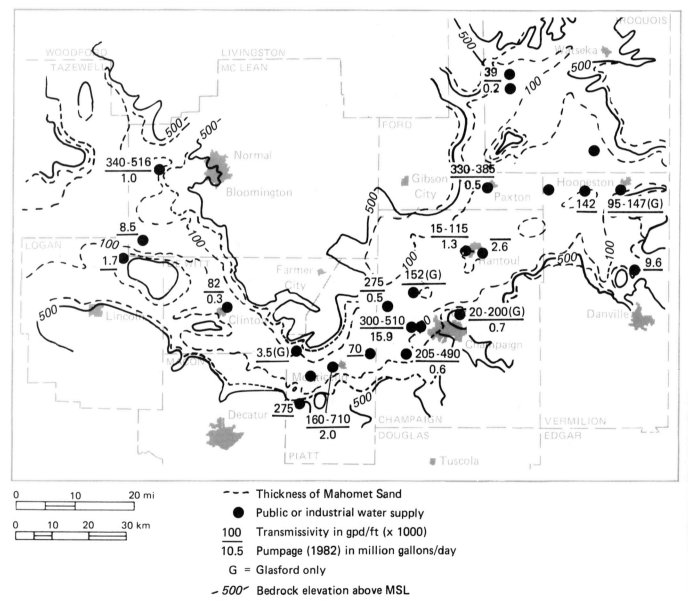

Figure 23. Representative hydrologic properties of municipal and industrial wells developed in the Mahomet Sand and Glasford Formation aquifer.

found in the region. Figure 24 shows the probable direction of the groundwater flow prior to extensive development of the Mahomet Sand as a water resource. The map is based on the reported water levels from the early wells drilled in the valley. The data are sparse and widely separated, and a precise potentiometric surface map could not be constructed.

The general pattern of groundwater flow, however, is clear from the available data. Water levels are highest in northwestern Vermilion County (Fig. 24), where the potentiometric head of the Mahomet Sand is almost as high as that of the Glasford Formation aquifer. Groundwater moves from areas of high to low total potentiometric head under the influence of gravity. The base levels, or discharge regions, appear to the major river valleys. Groundwater flows from the potentiometric high in northwest Vermilion County south and west, and ultimately discharges in the Illinois River Valley west of the junction with the Mackinaw Bedrock Valley. Groundwater flow is also likely northward, discharging to the Iroquois River Valley in central Iroquois County, and eastward, discharging to the Wabash River Valley in east-central Indiana.

Flowing wells occur where the potentiometric surface intersects or is above the land surface. This condition is commonly found in north-central Vermilion County in the valleys of the North and Middle Forks, Vermilion River, and in southern Iroquois County; flowing wells are occasionally found locally along the Sangamon River Valley in Champaign and Piatt Counties and some small valleys in north-central Ford and Iroquois Counties.

Less can be deduced about the groundwater flow systems in the Glasford Formation aquifers. Although the aquifers are almost all confined by younger Wisconsinan-age till, the aquifers do not appear to have major regional controls on the groundwater

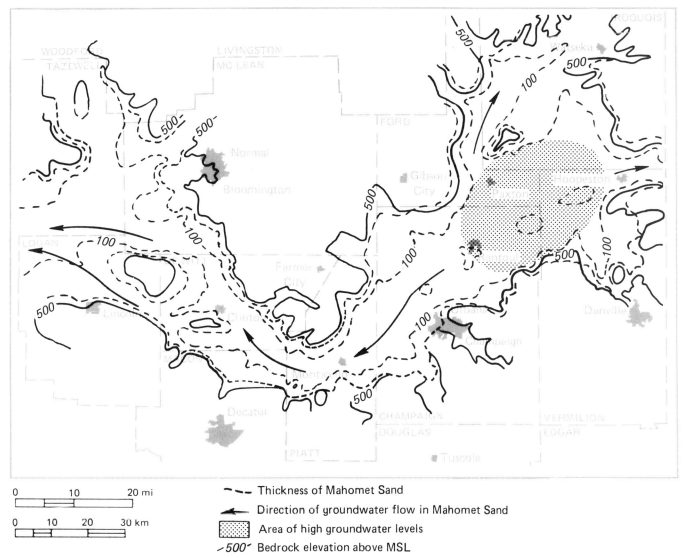

Figure 24. Areas of high groundwater levels and direction of regional groundwater flow in Mahomet Sand.

flow, suggesting that local or intermediate groundwater flow systems are dominant.

GENERAL SUMMARY AND CONCLUSIONS

While this study has been based primarily on a compilation of existing data and mapping, some reinterpretation of that data and additional mapping have aided in providing a more integrated characterization of the bedrock topography, valley-fill stratigraphy, Quaternary history, and hydrogeology of the Mahomet Valley Lowland. Many of the local bedrock "anomalies," seemingly unrelated stratigraphic sequences, conflicting interpretations of the geologic history, and fragmented hydrogeologic data now have begun to fit into a more consistent and understandable pattern.

The uneven distribution, limited availability, and in places, poor quality of data remain a significant limitation to specific interpretations. Although large numbers of water-well logs are available, the shallow depth of these wells, the often poor quality of the logs, the small number and questionable quality of the samples, and the questionable location of some of the wells limit their use. Relatively few wells penetrate the entire drift sequence to bedrock, particularly over the Mahomet Valley. A lack of representative sample through the entire drift sequence restricts the ability to do regional correlations or to adequately identify and characterize deeper drift units for large portions of the area. It has been possible to make some general lithic characterizations where both driller logs and sample cuttings of drift units are available and the drift-bedrock contact can be identified from downhole geophysical logs. However, resolution is often poor for the drift units. Given these limitations, it has been gratifying that a consistent regional pattern of the bedrock topography and general drift stratigraphy has emerged.

The probability that glaciation occurred throughout the Mahomet Valley Lowland prior to and during deposition of the main body of the Mahomet Sand in itself indicates a much longer and even more complex history of development of the Mahomet and associated bedrock valleys than has been previously suggested. The two principal schools of thought on the age of entrenchment of the major bedrock valleys—"preglacial" versus "deep" stage erosion during the Pleistocene—should now be re-evaluated in light of the new data from the Mahomet Bedrock Valley Lowland and associated valleys.

Horberg (1945, 1946, 1950), Horberg and Anderson (1956), and other workers have supported a preglacial development for most of the major bedrock valleys in the central United States during successive erosion or planation cycles, culminating in a deep bedrock valley stage. The preglacial surface was then only locally modified as a result of glaciation. Frye (1963) has summarized both points of view and suggested that there are significant data to support that the deepest erosion of the valleys did not take place until sometime well into the Pleistocene.

Evidence from the current study of the Mahomet Bedrock Valley Lowland and recent studies elsewhere provide some evidence that may help to resolve these conflicting views, at least within the Mahomet Valley Lowland. An analysis of the data and interpretations developed during this study can be separated into two categories: (1) the factors that controlled the development of the Mahomet Bedrock Valley Lowland, and (2) the relation of drift stratigraphy and distribution to the bedrock surfaces.

Bedrock valley development

The following are the principal features and their implications:

1. The major tributaries have a dendritic pattern, suggesting that they were part of a larger drainage system.

2. The pattern of tributaries in relation to the main Mahomet Valley Lowland in eastern Illinois suggests a headwaters region (e.g., Onarga Valley).

3. The transverse character of Mahomet Valley Lowland suggests that the ancestral drainage system developed on relatively flat-lying Pennsylvanian rocks and was superimposed across older bedrock structures.

4. The irregular nature of portions of the lowland--including benches, hills, and channels—suggests a complex erosional history.

5. The low gradient of the valley along its length suggests possible effect of both isostatic changes and short periods of rapid cutting.

6. The "hanging" tributary valleys suggest local rapid cutting of the main valley.

Relation of drift stratigraphy to bedrock surface

The following relations provide some insight into the development and relative age of the Mahomet Valley and the drift fill:

1. Glacial tills found on the bedrock below the Mahomet Sand at elevations above 450 ft (137 m), within the Mahomet Valley Lowland, indicate erosion of the lowland at least to this depth prior to local glaciation.

2. No deposits have been found on the bedrock floor of the lowland, indicating nonglacial alluvium. Therefore, there is no record of alluvial sedimentation in the lowland prior to mid-continent Quaternary glaciation.

3. Upland deposits adjacent to the Mahomet Lowland, lying directly on bedrock and below deposits correlated with pre-Illinoian sediments in the valley, are older than the Brunhes-Matuyama reversal, which suggests some of the deposits in the deeper part of the valley are older than about 730,000 B.P.

From the above evidence, we now believe that the Mahomet Valley Lowland developed through Late Tertiary time as a part of a westward-flowing drainage system with headwaters in northeastern Illinois and northwestern Indiana. This system was then disrupted by an early Pleistocene glaciation reaching near or into the drainage basin. Downcutting was accelerated either by development of a forebulge or by ice-margin channeling and/or ponding, resulting in episodes of "catastrophic" high-velocity floods, producing early Pleistocene downcutting and deepening of the lowland. Subsequent glaciers them moved across the valley, resulting in both till and outwash deposition.

Significance to groundwater resources

The widespread distribution of the thick, coarse-grained Mahomet Sand Member throughout the Mahomet Valley Lowland in central Illinois provides the source for the most productive wells in the region. Present groundwater pumpage is approaching 90 mgd (3×10^8 l/d); this may represent only a small portion of the total potential yield of the aquifers. An understanding of the geologic history of the development of this aquifer has provided a basis for predicting its hydrologic characteristics and regional groundwater availability. Additional, more detailed hydrogeologic investigations of the Mahomet Sand, the overlying Glasford aquifers, and the related confining beds will serve not only to better predict their performance as aquifers, but also will allow for a better understanding of the geologic framework and history.

ACKNOWLEDGMENTS

This chapter has benefited from contributions by many individuals. Particular thanks are due H. D. Glass, who provided x-ray diffraction analyses of core samples from numerous borings, and who has spent many hours over the years discussing the results and stratigraphic implications of the analyses with us. Particle-size analyses were provided by the Inter-Survey Geotechnical Laboratory under the supervision of Michael V. Miller. Adrian P. Visocky, Illinois State Water Survey, provided hydrologic data, discussed its use, and reviewed the manuscript. Robert C. Vaiden aided in compiling and plotting the basic data and in preparing all of the subsurface maps.

Ned K. Bleuer, Indiana Geological Survey, provided data on the Ambia test hole and has discussed stratigraphic relations between eastern Illinois and western Indiana with us over the past 20 yr. Ardith K. Hansel, Illinois Geological Survey, provided a critical review of the manuscript, as did Richard C. Anderson and David A. Stephenson, the official Geological Society of America reviewers. All three provided excellent comments and suggestions for which we are grateful.

Preparation of this manuscript, a consolidation of three papers presented at the 1983 Annual Meeting of the Geological Society of America, has been greatly aided by the patience and dedication of Jaquelyn L. Hannah (illustrations), and by Joanne Klitzing, Debra S. Sands, and Edna M. Yeargin (text), who prepared and revised numerous drafts of the evolving figures and manuscript over a 6-yr period.

REFERENCES CITED

Baker, V. R., 1978, Large-scale erosional and depositional features of the Channeled Scabland, *in* Baker, R. V., and Nummendal, D., eds., The Channeled Scabland: National Aeronautics and Space Administration, p. 81–115.

Bleuer, N. K., 1976, Remnant magnetism of Pleistocene sediments of Indiana: Indiana Academy of Science Proceedings, v. 85, p. 277–294.

——, 1983, Rock and event stratigraphy, *in* Bleuer, N. K., Melhorn, W. N., and Pavey, R. R., Interlobate stratigraphy of the Wabash Valley, Indiana: 30th Annual Midwest Friends of the Pleistocene Field Conference Guidebook, p. 21–28.

Boellstorff, J., 1978, North American Pleistocene Stages reconsidered in light of probable Pliocene-Pleistocene continental glaciation: Science, v. 202, p. 305–307.

Bredehoeft, J. D., 1957, Refraction seismic studies in the Havana Lowland area, Mason County, Illinois [M.S. thesis]: Urbana, University of Illinois, 33 p.

Bruns, T. M., Logan, S. M., and Steen, W. J., 1985, Map showing bedrock topography of the Teays Valley, western part, north-central Indiana: Indiana Geological Survey Miscellaneous Map 42, scale 1:100,000.

Buhle, M. B., and Brueckmann, J. E., 1964, Electrical earth resistivity surveying in Illinois: Illinois State Geological Survey Circular 376, 51 p.

Cartwright, K., 1972, Bedrock topography of east-central Illinois: Illinois State Geological Survey Open-File Map (unpublished) scale 1:250,000.

Clegg, K. E., 1972, Subsurface geology and coal resources of the Pennsylvanian System in De Witt, McLean, and Piatt Counties, Illinois: Illinois State Geological Survey Circular 473, 25 p.

Fidler, M. M., 1943, The preglacial Teays Valley in Indiana: Journal of Geology, v. 51, p. 411–418.

Follmer, L. R., 1983, Sangamon and Wisconsinan pedogenesis in the Midwestern United States, *in* Porter, S. C., ed., The Late Pleistocene, v. 1 of Wright, H. E., ed., Late Quaternary environments of the United States: Minneapolis, University of Minnesota Press, p. 138–144.

Ford, J. P., 1974, Surficial deposits in Coles County, Illinois: Illinois Geological Survey Open-File Manuscript.

Foster, J. W., 1952, Major aquifers in glacial drift near Mattoon, Illinois: Illinois Academy of Science Transactions, v. 44, p. 85–94.

——, 1953, Significance of Pleistocene deposits in the groundwater resources of Illinois: Economic Geology, v. 48, no. 7, p. 568–573.

Foster, J. W., and Buhle, M. B., 1951, An integrated geophysical and geological investigation of aquifers in glacial drift near Champaign-Urbana, Illinois: Economic Geology, v. 46, no. 4, p. 367–397; Illinois State Geological Survey Report of Investigations 155, 31 p.

Frye, J. C., 1963, Problems of interpreting the bedrock surface of Illinois: Illinois Academy of Science Transactions, v. 56, no. 1, p. 3–11.

Fullerton, D. S., 1986, Stratigraphy and correlation of glacial deposits from Indiana to New York and New Jersey, *in* Sibrava, V., Bowen, D. Q., and Richmond, G. M., eds., Quaternary glaciations in the Northern Hemisphere: Quaternary Science Reviews, v. 5, p. 23–37.

Gibb, J. P., 1970, Groundwater availability in Ford County: Illinois Water Survey Circular 97, 66 p.

Glass, H. D., Frye, J. C., and Willman, H. B., 1964, Record of diversion in the Morton Loess of Illinois: Illinois Academy of Science Transactions, v. 57, no. 1, p. 24–27.

Gray, H. H., 1982, Map of Indiana showing topography of the bedrock surface: Indiana Geological Survey Miscellaneous Map 35, scale 1:500,000.

Hallberg, G. R., 1980, Pleistocene stratigraphy in east-central Iowa: Iowa Geological Survey Technical Information Series 10, 168 p.

——, 1986, Pre-Wisconsinan glacial stratigraphy of central plains regions in Iowa, Nebraska, Kansas, and Missouri, *in* Sibrava, V., Bower, D. Q., and Richmond, G. M., eds., Quaternary glaciations in the Northern Hemisphere: Quaternary Science Reviews, v. 5, p. 11–15.

Heigold, P. C., and Ringler, R. W., 1979, A seismic refraction survey of the lower Illinois Valley bottomlands: Illinois State Geological Survey Circular 507, 18 p.

Heigold, P. C., McGinnis, L. D., and Howard, R. H., 1964, Geologic significance of the gravity field in the De Witt–McLean County area, Illinois: Illinois State Geological Survey Circular 369, 16 p.

Horberg, L., 1945, A major buried valley in east-central Illinois and its regional relationships: Journal of Geology, v. 53, no. 5, p. 349–359; Illinois State Geological Survey Report of Investigations 106, 11 p.

——, 1950, Bedrock topography of Illinois: Illinois State Geological Survey Bulletin 73, 111 p.

——, 1953, Pleistocene deposits below the Wisconsin drift in northeastern Illinois: Illinois State Geological Survey Report of Investigations 165, 61 p.

——, 1956, Preglacial erosion surfaces in Illinois: Journal of Geology, v. 54, no. 3, p. 179–192.

Horberg, L., and Anderson, R. C., 1956, Bedrock topography and Pleistocene glacial lobes in central United States: Journal of Geology, v. 64, no. 2, p. 101–106.

Horberg, L., Suter, M., and Larson, T. E., 1950, Groundwater in the Peoria region: Illinois State Geological Survey Bulletin 75, 128 p.

Hunt, C. S., and Kempton, J. P., 1977, Geology for planning in De Witt County, Illinois: Illinois State Geological Survey Environmental Geology Notes 83, 42 p.

Jacobs, A. M., and Lineback, J. A., 1969, Glacial geology of the Vandalia, Illinois, region: Illinois State Geological Survey Circular 442, 23 p.

Johnson, W. H., 1964, Stratigraphy and petrography of Illinoian and Kansas drift in central Illinois: Illinois State Geological Survey Circular 378, 38 p.

——, 1971, Old glacial drift near Danville, Illinois: Illinois State Geological Survey Circular 457, 16 p.

——, 1976, Quaternary stratigraphy in Illinois; Status and current problems, in Mahoney, W. C., ed., Quaternary stratigraphy of North America: Stroudsburg, Pennsylvania, Dowden Hutchison and Ross, p. 161–196.

——, 1983, Sedimentological interpretation of the Belgium (Hegeler) Member, Banner Formation, eastern Illinois: Geological Society of America Abstracts with Programs, v. 15, p. 224.

——, 1986, Stratigraphy and correlation of the glacial deposits of the Lake Michigan Lobe prior to 14 ka B.P., in Sibrava, V., Bowen, D. Q., and Richmond, G. M., eds., Quaternary glaciations in the Northern Hemisphere: Quaternary Science Reviews, v. 5, p. 17–22.

Johnson, W. H., Gross, D. L., and Moran, S. R., 1971, Till stratigraphy of the Danville region, Illinois, in Goldthwait, R. P., and others, eds., Till; A symposium: Columbus, Ohio State University Press, p. 184–216.

Johnson, W. H., Follmer, L. R., Gross, D. L., and Jacobs, A. M., 1972, Pleistocene stratigraphy of east-central Illinois: Illinois State Geological Survey Guidebook Series 9, 97 p.

Kehew, A. E., and Lord, M. L., 1986, Origin and large-scale erosional features of glacial-lake spillways in the northern Great Plains: Geological Society of America Bulletin, v. 97, p. 162–177.

——, 1988, Glacial lake outbursts along the midcontinent margins of the Laurentide ice sheet, in Mayer, L., and Nash, D., eds., Catastrophic flooding: Binghamton Symposium in Geomorphology, no. 18, p. 95–120.

Kempton, J. P., and Berg, R. C., 1982, Bedrock topography of Kankakee County, Illinois: Illinois State Geological Survey Open-File Map (unpublished), scale 1:62,500.

Kempton, J. P., and Hackett, J. E., 1962, Use of physical properties in subsurface studies of glacial materials [abs.]: Geological Society of America Special Paper 68, p. 210.

——, 1968, Stratigraphy of the Woodfordian and Altonian drifts in central-northern Illinois, in Bergstrom, R. E., ed., The Quaternary of Illinois: Urbana, University of Illinois College of Agriculture Special Publication 14, p. 27–34.

Kempton, J. P., and Reed, P. C., 1973, Wider, shallower Paw Paw bedrock valley and complex drift sequence revealed by drilling: Geological Society of America Abstracts with Programs, v. 5, p. 325–326.

Kempton, J. P., Morse, W. J., and Johnson, W. H., 1980, Stratigraphy and regional distribution of Illinoian and pre-Illinoian deposits of east-central Illinois: Geological Society of America Abstracts with Programs, v. 12, p. 230–231.

Kempton, J. P., Ringler, R. W., Heigold, P. C., Cartwright, K., and Poole, V. L., 1981, Groundwater resources of northern Vermilion County, Illinois: Environmental Geology Notes 101, 36 p.

Kempton, J. P., Morse, W. J., and Visocky, A. P., 1982, Hydrogeologic evaluation of sand and gravel aquifers for municipal groundwater supplies in east-central Illinois: Illinois State Geological Survey and Illinois State Water Survey Cooperative Groundwater Report 8, 59 p.

Killey, M. M., and Lineback, J. A., 1983, Stratigraphic reassignment of the Hagarstown Member in Illinois, in Geologic notes: Illinois State Geological Survey Circular 529, p. 13–16.

Leighton, M. M., 1959, Stagnancy of the Illinoian glacial lake east of the Illinois and Mississippi Rivers: Journal of Geology, v. 67, no. 3, p. 337–344.

Leighton, M. M., and Brophy, J. A., 1961, Illinoian glaciation in Illinois: Journal of Geology, v. 69, no. 1, p. 1–31.

Manos, C., 1961, Petrography of the Teays–Mahomet Valley deposits: Journal of Sedimentary Petrology, v. 31, no. 3, p. 456–460.

McGinnis, L. D., 1968, Glacial crustal bending: Geological Society of America Bulletin, v. 79, p. 769–775.

McGinnis, L. D., and Heigold, P. C., 1974, A seismic refraction survey of the Meredosia channel area of northwestern Illinois: Illinois State Geological Survey Circular 488, 19 p.

McGinnis, L. D., Kempton, J. P., and Heigold, P. C., 1963, Relationship of gravity anomalies to a drift-filled bedrock valley system in northern Illinois: Illinois State Geological Survey Circular 354, 23 p.

Nelson, R. S., 1981, Bedrock topography of McLean County: Illinois State Geological Survey Open-File Map (unpublished), scale 1:250,000.

Piskin, K., and Bergstrom, R. E., 1975, Glacial drift in Illinois; Thickness and character: Illinois State Geological Survey Circular 490, 35 p.

Reinertsen, D. L., Berggren, D. J., Kempton, J. P., and DuMontelle, P. B., 1977, Champaign-Urbana area: Illinois Geological Survey Earth Science Guide Leaflet 1977B.

Richards, S. S., and Visocky, P. A., 1982, A reassessment of aquifer conditions west of Normal, Illinois: Illinois State Water Survey Circular 153, 33 p.

Sanderson, E. W., 1971, Groundwater availability in Piatt County: Illinois State Water Survey Circular 107, 83 p.

Selkregg, L. F., and Kempton, J. P., 1958, Groundwater geology in east-central Illinois; A preliminary geologic report: Illinois State Geological Survey Circular 248, 36 p.

Stephenson, D. A., 1967, Hydrogeology of glacial deposits of the Mahomet Bedrock Valley in east-central Illinois: Illinois State Geological Survey Circular 409, 51 p.

Stout, W. E., Ver Steeg, K., and Lamb, G. F., 1943, Geology of water in Ohio: Ohio Geological Survey, 4th series, Bulletin 44, 694 p.

Teller, J. T., and Thorleifson, L. H., 1983, The Lake Agassiz–Lake Superior connection, in Teller, J. T., and Clayton, L., eds., Glacial Lake Agassiz: Geological Association of Canada Special Paper 26, p. 260–290.

Visocky, A. P., and Schicht, R. J., 1969, Groundwater resources of the buried Mahomet Bedrock Valley: Illinois State Water Survey Report of Investigations 62, 52 p.

Walker, W. H., Bergstrom, R. E., and Walton, W. C., 1965, Preliminary report on the ground-water resources of the Havana region in west-central Illinois: Illinois State Water Survey and Geological Survey Cooperative Ground-Water Report 3, 61 p.

Wayne, W. J., 1956, Thickness of drift and bedrock physiography of Indiana north of the Wisconsin glacial boundary: Indiana Geological Survey Report of Progress 7, 70 p.

Willman, H. B., and Frye, J. C., 1970, Pleistocene stratigraphy of Illinois: Illinois State Geological Survey Bulletin 94, 204 p.

Willman, H. B., and others, 1967, Geologic map of Illinois: Illinois State Geological Survey, scale 1:500,000.

Woller, D. M., 1975, Public groundwater supplies in Champaign County: Illinois State Water Survey Bulletin 60-15, 55 p.

MANUSCRIPT ACCEPTED BY THE SOCIETY JUNE 29, 1990

Printed in U.S.A.

The Teays System; A summary

Wilton N. Melhorn
Department of Earth and Atmospheric Sciences, Purdue University, West Lafayette, Indiana 47907
John P. Kempton
Illinois State Geological Survey, Natural Resources Building, 615 East Peabody Drive, Champaign, Illinois 61820

We now have read a new series of chapters devoted to an old topic, for glacially diverted streams and buried ancestral valleys were discovered in the mid-continental United States more than a century ago. For example, there are the papers by Bradley (1870, p. 230) on the Kankakee district in the Worthen Survey reports (1866 to 1875) of Illinois, and by Brown (1882) describing buried valleys in Fountain County, western Indiana, in the Collett Survey reports. The first areal or regional syntheses come from Leverett (1895), in a discussion of the buried Ancient Mississippi River Valley; by Bownocker (1899), for western Ohio and eastern Indiana; from Leverett in the classic Monograph 38 (1899); and of course, the famous reconstruction and naming of the Teays River System by Tight (1903).

Advances in the geological sciences often seem to come in spurts and surges. Four decades passed after the pioneering works before another gaggle of studies emerged; for example, the work of Stout and others in Ohio (1943), Fidlar (1943) in Indiana, and Horberg (1945) in Illinois. All of these acknowledgeably were spawned by the exigencies of World War II, specifically the need to identify sources and abundance of groundwater in support of the war-effort requirements of Smokestack America. Now, after another four decades, again we enter a time of generally predicted groundwater shortages and recognition of the need to protect the quality of this resource. Thus it is fitting that some chapters in this volume deal either directly or obliquely with the water resources and formation characteristics of aquifers within the Teays-Mahomet System.

It is inappropriate, and indeed there is no space in a summary, to recapitulate all details contained in the several symposium papers, to discourse on points of agreement or disagreement among the authors, or to unduly emphasize some papers at the expense of others. Judgments are best left to the reader. However, by rushing in where even angels fear to tread, it seems that two points emerge: that (1) a common theme permeates most of these chapters, and that (2) the classically identified, preglacial, integrated Teays-Mahomet valley system is incorrect, or at least in extreme doubt, as a result of a reinterpretation of currently available evidence.

With respect to the first point, all chapters, at least inferentially, expectably emphasize and deal with the matter of assembling more and better subsurface data about the Teays. Included are such items as more precise topographic and subsurface elevational controls; reliable seismic, resistivity, and possibly gravity data; and an increased use of modern, more sophisticated methods of drilling, sampling, and stratigraphic interpretation. Likewise important—and formerly not readily available—are paleomagnetic, paleoclimatic, and soils information; restoration of former erosion and weathering surfaces; and clay, textural, and heavy-mineral analyses. At this point, however, feedback or cause-and-effect relations becomes important. Continuous acquisition of more and better basic data clearly is a forward step in the resolution of areal questions about the age, course, and history of a "Teays River." Thus, as former shadows of doubt fade, assuredly even better methods will emerge for acquiring geologic and hydrologic data, and in turn will provide additional feedback to further dispel the mystery of the origin, history, and ages of various riverine links and segments that seem to actually make up the "Teays System." Perhaps we shall not have to wait another 40 years for the next symposium!

As to the second point, at least in the traditional sense, as shown for example, in Flint (1971, p. 235), was there ever *really* a "Teays"? Or, as stated in this volume by Henry Gray, who obviously paraphrases Gertrude Stein, ". . . we too easily accept that the Teays is the Teays is the Teays." Assuredly, anyone who learned well their lessons in basic physical geology and geomorphology would see that the "classic Teays" of most maps is irrational, for the projected flow path defies the general rule that in a relatively stable tectonic setting (such as the Midwest), major streams should adjust to structure, lithology, and existing topography. Yet the "classic Teays" does none of these, and from the chapters in this volume, it seems reasonable to conclude that the "Teays likely is *not* the Teays," either in the sense of a discrete river or a single, linear, continuous valley as envisioned by the early workers, long accepted as dogma by standardized geology texts, or as seen in popular articles exemplified by yellowed newspaper clippings that litter our file drawers. Stated another way, in the immortal words of Spock: "This does not compute, Captain." But there *is* a Teays System in the sense that we conven-

Melhorn, W. N., and Kempton, J. P., 1991, The Teays System; A summary, *in* Melhorn, W. N., and Kempton, J. P., eds., Geology and hydrogeology of the Teays-Mahomet Bedrock Valley System: Boulder, Colorado, Geological Society of America Special Paper 258.

ors, in blissful ignorance, originally chose as a title for the symposium, though in reality it seems that, taken in the entirety, what most geologic maps show as a continuous, buried valley is an illusion, and there exists the remnant elements of at least three "Teays rivers."

First, there is an "eastern Teays," which likely existed in late Tertiary time, and discharged northward, by some yet undetermined route, to the general area of the Erie Lowland. This is not a new idea; the logic was seen by Coffey (1958), and despite failures to date, the exit may yet be found. Perhaps the relatively unexplored nuances of "glacial forebulge" and other isostatic revelations cloud our thinking. This is not a dead issue in Ohio, where some workers continue to seek the "lost" routing.

Second, there is a "western Teays" or Mahomet drainage, likely but not conclusively demonstrated as also of at least late Tertiary age. One headwater branch was in extreme eastern Illinois, or more likely on the Norman Upland of west-central Indiana, on the dip slope of the Knobstone (Borden) escarpment. Perhaps by late pre-Illinoian (i.e., Kansan) time or earlier, this Mahomet drainage had breached the escarpment to capture a

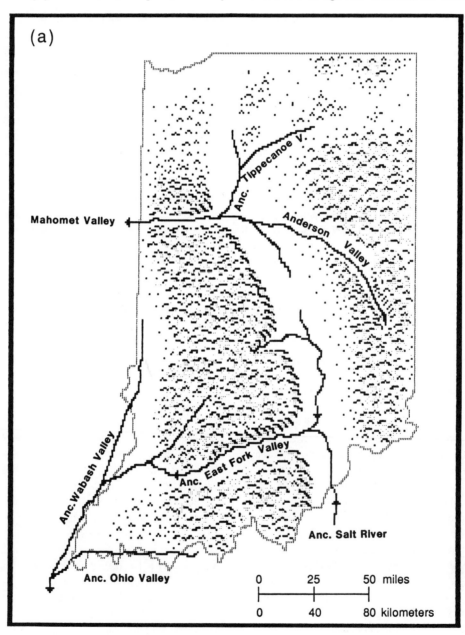

Figure 1a. A hypothesized, preglacial stage of drainages in Indiana. The Mahomet System drains much of north-central Indiana east of the Knobstone Escarpment as a result of stream capture of the Lafayette Lowland. No drainage is shown for extreme northern Indiana, but the area probably contained short streams that flowed to the Michigan (and Erie?) lowlands. Neither the present White or Ohio Rivers have yet fully developed.

northward-flowing drainage of the Lafayette Lowland to the east (Fig. 1a). In any event, other evidence (width, depth, gradients, orientation of tributaries, etc.) suggests that the Teays of Tight and the Mahomet of Horberg truly were separate entities.

Third and finally, there may have been one or more "glacial-age Teays" that flowed across much of north-central and eastern Indiana, and probably western Ohio as well, each channel briefly existing as a member of a shifting plexus created by episodic diversions, captures, pondings, or overflows across minor divides. For example, one possibility is hypothesized in Figure 1b, wherein pre-Illinoian (i.e., Kansan) ice from the northeast (perhaps a pre-Wilshire, pre–West Labanon event?) blocks existing local, northward-flowing drainages (to the Michigan Lowland?). In this imaginative but plausible scenario, ice-margin drainage is diverted westward across north-central Indiana into the Mahomet headwaters west of Lafayette. Other scenarios could be devised.

As already noted, what previously was known about this

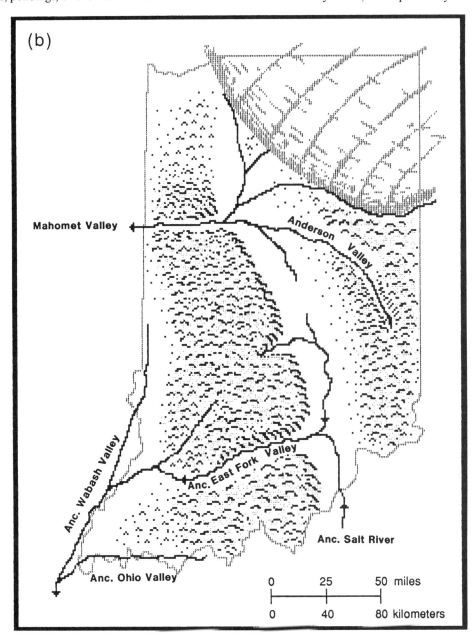

Figure 1b. An early glacial advance from the northeast, probably prior to 700,000 B.P. (i.e., Kansan) blocks northward-flowing Teays (and other) drainages in northwestern Ohio and diverts meltwaters across Indiana as ice-marginal flow, cutting a deep, narrow valley gorge in the eastern part of the state. This outflow, whether a single event or as a series of jokulhlaups, crosses the Lafayette Lowland to join the Mahomet drainage. Whatever glacier created this valley antedates the >0.7-Ma West Lebanon event, which filled the valley.

buried valley system came from interpretations of drilling data obtained during sporadic, regional hydrologic investigations. These studies revealed long ago that the western, or Mahomet reach, either yields or has the capacity to yield, significant quantities of potable groundwater. Results of water-supply investigations in the buried valley of western Ohio are somewhat mixed, but generally disappointing. This volume shows, for the first time, that from Lafayette to Marion in Indiana, stratigraphic and lithologic changes in valley fill generally indicate a progressive eastward deterioration in aquifer thickness and lateral extent. This is owed to historical constraints imposed by the existence of barrier boundaries, plugging of valley fill with fine-grained, ponded sediments, or mixing of unsorted, ice-front flow tills with sorted, proglacial outwash. Generally, also, the wider the trunk valley, the thicker and broader the aquifer sequence; where there is a narrow, gorge-like valley as in the Marion segment, aquifers are thinner and less broadly distributed.

There remain, of course, questions about demonstrated or hypothesized flat or reversed buried valley-bottom gradients, a supposed lack of karst solutioning that presumably should exist on certain buried, bedrock uplands, the apparent absence of dissection by streams tributary to the master channel, and other matters that crop up in the literature as having puzzled earlier researchers. All these are yet to be resolved, as is how much the true picture is complicated by forebulge effects, multiple ice dams, captures and piracy, or some combination of these and other unknown phenomena. Regardless, it seems to us that the buried "Teays Valley," as it appears on even the most modern of maps (Gray, 1982; Bruns and others, 1985) is more an artifact of map-making than an appeal to reality, and that the "Teays" consists of varied segments that neither are coeval in time nor continuous and linear in geometry.

ACKNOWLEDGMENTS

The following geologists served as reviewers of symposium manuscripts appearing in this volume: R. C. Anderson, D. L. Edgar, L. R. Follmer, J. L. Forsyth, D. L. Gross, J. C. Knox, D. I. Leap, L. D. McGinnis, D. M. Mickelson, A. F. Schneider, D. A. Stephenson, W. T. Straw, S. M. Totten, and W. J. Wayne. We gratefully acknowledge these timely and thorough reviews.

REFERENCES CITED

Bownocker, J. A., 1899, A deep preglacial channel in western Ohio and eastern Indiana: American Geologist, v. 23, p. 178–182.

Bradley, F. H., 1870, Geology of Kankakee and Iroquois Counties, in Worthen, A. H., Geological Survey of Illinois: Geological Survey of Illinois, v. 5, p. 226–240.

Brown, R. T., 1882, Fountain County; Geology, geography, etc., in Department of Geology and Natural Resources 11th Annual Report for the year 1881: Indianapolis, Indiana, Wm. Burford, p. 84–125.

Bruns, T. M., Logan, S M., and Steen, W. J., 1985, Map showing bedrock topography of the Teays Valley: Indiana Geological Survey Miscellaneous Map 42, 3 sheets, scale 1:100,000.

Coffey, G. N., 1958, Major glacial drainage changes in Ohio: Ohio Journal of Science, v. 58, no. 1, p. 43–49.

Fidlar, M. M., 1943, The preglacial Teays Valley in Indiana: Journal of Geology, v. 51, p. 411–418.

Flint, R. F., 1971, Glacial and Pleistocene geology: New York, John Wiley & Sons, 892 p.

Gray, H. H., 1982, Map of Indiana showing topography of the bedrock surface: Indiana Geological Survey Miscellaneous Map 36, 1 sheet, scale 1:500,000.

Horberg, C. L., 1945, A major buried valley in east-central Illinois and its regional relationships: Journal of Geology, v. 53, p. 349–359.

Leverett, F., 1895, The preglacial valley of the Mississippi and its tributaries: Journal of Geology, v. 3, p. 740–763.

—— , 1899, The Illinois glacial lobe: U.S. Geological Survey Monograph 38, 817 p.

Stout, W. E., Ver Steeg, K., and Lamb, G. F., 1943, Geology of water in Ohio: Ohio Geological Survey, 4th series, Bulletin 44, 694 p.

Tight, W. G., 1903, Drainage modifications in southeastern Ohio and adjacent parts of West Virginia and Kentucky: U.S. Geological Survey Professional Paper 13, 111 p.

MANUSCRIPT ACCEPTED BY THE SOCIETY JUNE 29, 1990

Typeset by WESType Publishing Services, Inc., Boulder, Colorado
Printed in U.S.A. by Malloy Lithographing, Inc., Ann Arbor, Michigan